Higher
BIOLOGY

The Scottish Certificate of Education Examination Papers
are reprinted by special permission of
THE SCOTTISH QUALIFICATIONS AUTHORITY

ISBN 0 7169 9307 4

ROBERT GIBSON · Publisher
17 Fitzroy Place, Glasgow, G3 7SF.

CONTENTS

SCOTTISH
CERTIFICATE OF
EDUCATION
1995

THURSDAY, 11 MAY
9.30 AM – 12.00 NOON

BIOLOGY
HIGHER GRADE
Paper II

INSTRUCTIONS TO CANDIDATES

1. All questions should be attempted: it should be noted however that questions in Section C each contain a choice.

2. The questions may be answered in any order but all answers are to be written in the spaces provided in this answer book, and must be written clearly and legibly in ink.

3. Additional space for answers and rough work will be found at the end of the book. If further space is required, supplementary sheets may be obtained from the Invigilator and should be inserted inside the <u>front</u> cover of this booklet.

4. The number of questions must be clearly inserted with any answers written in the additional space.

5. Rough work, if any should be necessary, should be written in this booklet and then scored through when the fair copy has been written.

6. Before leaving the examination room this book must be given to the Invigilator. If you do not, you may lose all the marks for this paper.

SECTION A

Answer ALL questions in this section.

Marks

1. (*a*) The diagram below represents part of a molecule of DNA on which a molecule of m-RNA is being synthesised.

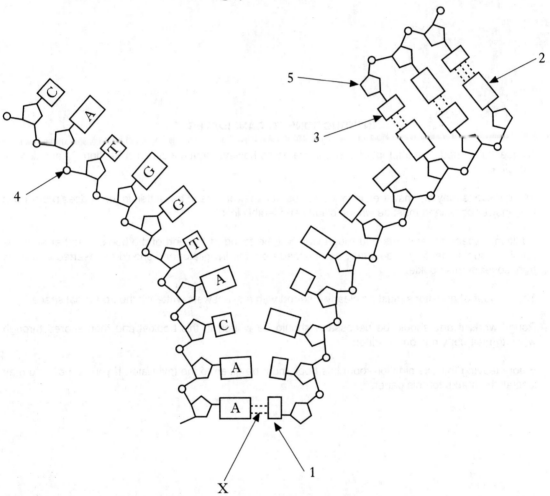

(i) From the diagram, identify bases 1, 2 and 3.

1. _____

2. _____

3. _____

(3)

Marks

(ii) Name the type of bond labelled as X in the diagram.

Type of bond _____ **(1)**

(iii) Name components labelled 4 and 5 in the diagram.

4. _____

5. _____ **(2)**

(b) In a DNA molecule, the base sequence AGT codes for the amino acid serine.

Using the initial letters of the bases, write the base sequence of the anti-codon on the t-RNA molecule to which serine becomes attached.

Space for working

Anti-codon _____ **(1)**

(c) The table below refers to features of the nucleic acids present in a human cell.

Place ticks (✓) in the appropriate boxes to indicate which of the statements are true for DNA and which are true for m-RNA.

Statement	DNA	m-RNA
Made in the nucleus		
Forms genes		
Attaches to ribosomes		

(2)

2. The diagram below shows a highly magnified view of part of a muscle cell.

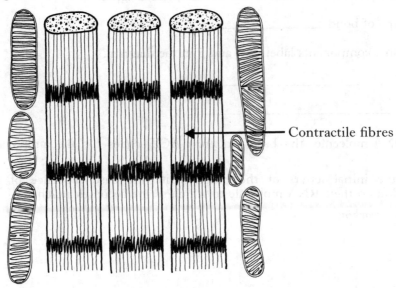

Contractile fibres

The flow diagram below summarises part of the process of respiration in the muscle cell.

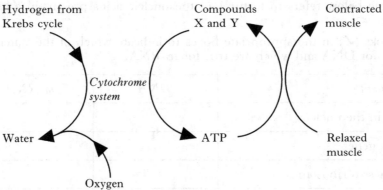

(*a*) What evidence, illustrated in the diagram, indicates that contractile fibres have a high energy demand?

_____

_____ **(1)**

(*b*) How is the hydrogen transferred from the Krebs cycle to the cytochrome system?

_____

_____ **(1)**

(*c*) Name compounds X and Y.

Compound X _____

Compound Y _____ **(1)**

Marks

(d) The list below contains information about respiration in muscle cells.

Information	Letter
Occurs in cytoplasm	A
Hydrogen released	B
Carbon dioxide released	C
Occurs anaerobically	D

Insert the appropriate letter in each of the boxes below to match the information to the stages of respiration.

One box has been completed for you.

Each letter can be used **once**, **more than once** or **not at all**.

Stage of respiration		Boxes
Glycolysis	→	A
Krebs cycle	→	
Conversion of pyruvic acid to lactic acid	→	

(3)

(e) The diagram below shows the arrangement of proteins in part of a cell membrane.

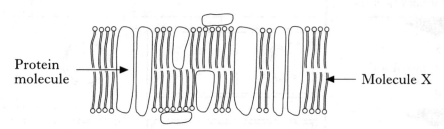

Protein molecule →   ← Molecule X

(i) State **one** property of the membrane which results from the arrangement of proteins in the membrane.

_____

_____ **(1)**

(ii) Name molecule X.

Name _____ **(1)**

3. Bacteria may be "engineered" to synthesise substances which are useful to humans. Human Growth Hormone may be produced in this way.

The diagrams below show stages in this process.

The plasmids contain a gene that gives resistance to an antibiotic called ampicillin.

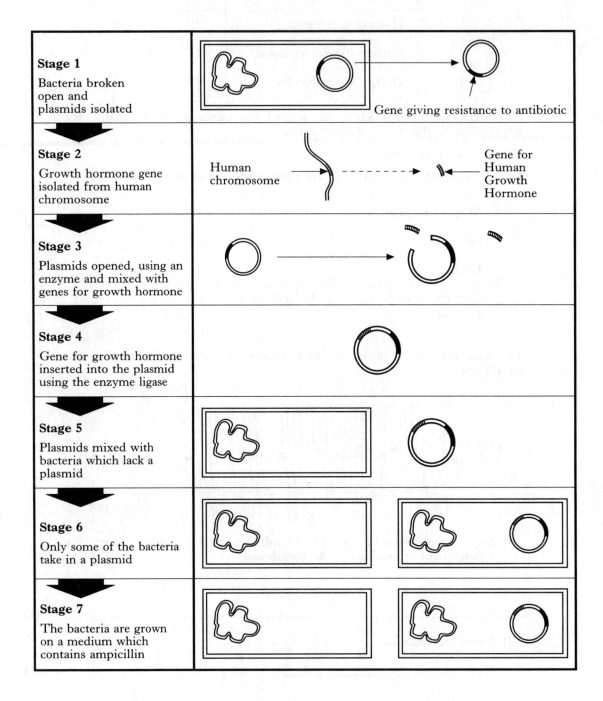

Stage 1 Bacteria broken open and plasmids isolated	Gene giving resistance to antibiotic
Stage 2 Growth hormone gene isolated from human chromosome	Human chromosome → → Gene for Human Growth Hormone
Stage 3 Plasmids opened, using an enzyme and mixed with genes for growth hormone	
Stage 4 Gene for growth hormone inserted into the plasmid using the enzyme ligase	
Stage 5 Plasmids mixed with bacteria which lack a plasmid	
Stage 6 Only some of the bacteria take in a plasmid	
Stage 7 The bacteria are grown on a medium which contains ampicillin	

Marks

(a) Name the enzyme which is used in Stage 3.

_____ **(1)**

(b) Give **one** advantage of the use of genetic engineering in the production of substances such as Human Growth Hormone.

_____

_____ **(1)**

(c) How does the addition of ampicillin to the medium at Stage 7 improve the yield of Human Growth Hormone in this process?

_____

_____

_____

_____ **(2)**

Marks

4. (*a*) The diagram below represents two responses of the human body to invasion by bacteria.

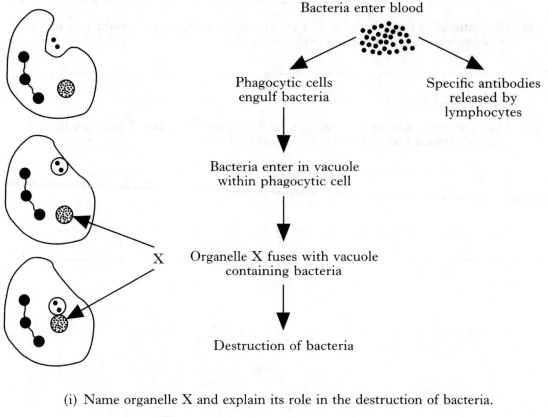

(i) Name organelle X and explain its role in the destruction of bacteria.

Name _____ **(1)**

Explanation _____

_____ **(1)**

(ii) Explain what is meant by the term "specific" with reference to antibodies.

_____

_____

_____ **(1)**

(*b*) Name one defence mechanism by which plants respond to invasion by micro-organisms.

_____

_____ **(1)**

Marks

5. The fruitfly, *Drosophila*, is used in genetic crosses.

Males are XY and females are XX.

In *Drosophila*, the gene for eye colour is sex-linked and red-eyed (R) is dominant to white-eyed (r).

Crosses between three different sets of parents were carried out as shown in the table below.

The numbers and phenotypes of the offspring obtained are also shown in the table.

Cross	Phenotypes of parents	Numbers and phenotypes of offspring			
		Red-eyed male	*Red-eyed female*	*White-eyed male*	*White-eyed female*
1	Red-eyed × Red-eyed male female	18	40	19	0
2	White-eyed × Red-eyed male female	17	18	19	18
3	Red-eyed × White-eyed male female	0	39	40	0

(a) Identify the genotypes of the parents in each of the above crosses.

Cross	Male genotype	Female genotype
1		
2		
3		

(3)

Space for working

(b) Which cross allows the sex of the offspring to be determined by their eye-colour?

Cross: _____

(1)

Marks

6. The myxoma virus kills rabbits.

Rabbit populations, in two areas (A and B), were exposed to the virus in an attempt to reduce the sizes of the populations.

The percentages of rabbits that survived each of a series of six successive epidemics are given in the table.

Epidemic number	Percentage of rabbits surviving	
	Area A	Area B
1	5	6
2	12	18
3	15	30
4	26	15
5	40	16
6	52	35

(a) How do the data for Area A support the statement that there has been selection for a resistant strain of rabbit?

_____

_____ **(1)**

(b) Give one reason why the first epidemic did not kill all rabbits exposed to the virus.

_____

_____ **(1)**

(c) What evidence is there from the data that the virus may have mutated?

_____

_____ **(1)**

Marks

7. (*a*) Plant tissue culture is a technique used by horticulturists for the production of many individual plants from a single parent plant.

Stages in the technique are shown below.

1. Removal of shoot tip from parent plant

2. Shoot tip transferred to multiplication medium

3. Mass of undifferentiated cells formed

4. Cells differentiate to form shoots

5. Separation of shoots

6. Individual shoots transferred to rooting medium

7. Small plants transferred to soil

(i) Explain why the shoot tip is suitable for use in this technique.

_____

_____ **(1)**

(ii) Which plant growth substance would be present in the rooting medium?

_____ **(1)**

(iii) Explain why the small plants produced can be described as being members of the same clone.

_____ **(1)**

(*b*) Give **one** reason why iron and calcium are essential for healthy growth and development in humans.

1. Iron: _____

_____ **(1)**

2. Calcium: _____

_____ **(1)**

13

8. Shrews are small mammals.

Marks

The graph below shows the relationship between oxygen consumption and body mass in shrews at two environmental temperatures.

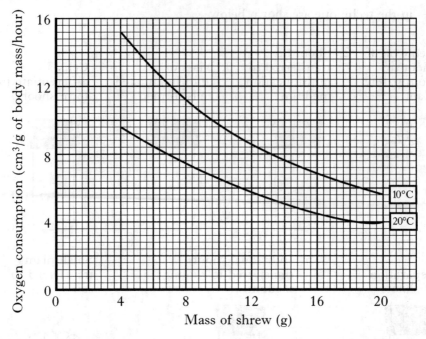

(*a*) Calculate the volume of oxygen used in two hours at 20 °C by a shrew of mass 20 g.

Space for calculation

Volume _____ cm^3 **(1)**

(*b*) State the relationship between the body mass of shrews and oxygen consumption.

_____

_____   **(1)**

(*c*) The shrew is an endotherm. Explain the meaning of this term.

_____

_____   **(1)**

(*d*) Explain the difference in oxygen consumption at 10 °C and at 20 °C by a shrew of mass 20 g.

_____

_____

_____   **(2)**

(*e*) Name the part of the brain which detects changes in the temperature of the blood.

_____   **(1)**

Marks

9. The flow diagram below outlines the self-regulating mechanism which controls the size of a population of mice in an ecosystem.

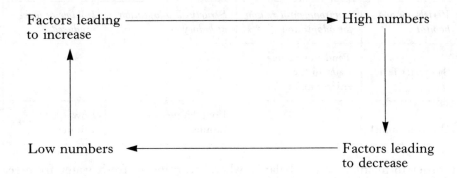

(*a*) State one way in which the foraging behaviour of predators may change as the number of mice decreases.

_____

_____ **(1)**

(*b*) List two **density-dependent** factors, other than predation, which could lead to a change in the size of the population of mice.

1. _____ **(1)**

2. _____ **(1)**

(*c*) Name one **density-independent** factor which could lead to a decrease in the size of the population of mice.

_____ **(1)**

Marks

10. (*a*) Complete the table below which shows adaptations for maintaining water balance in two vertebrates from different habitats.

Vertebrate and habitat	Environmental reason for adaptation	Structural adaptation of kidney	Volume of urine produced
Salt water fish	Tendency to lose water to the environment		
Desert mammal		Long kidney tubules	Low volume

(3)

(*b*) In order to maintain water balance when returning to fresh water from sea water, the Atlantic salmon shows adaptive changes.

Underline one word in each group to complete the sentences correctly.

As a result of these adaptive changes, the rate of glomerular filtration $\left\{ \begin{array}{l} \text{increases} \\ \text{decreases} \end{array} \right\}$.

Specialised cells in the gills $\left\{ \begin{array}{l} \text{increase} \\ \text{decrease} \end{array} \right\}$ the rate of salt uptake from the environment.

(1)

Marks

(*c*) The diagram below shows the section through the floating leaf of a hydrophyte.

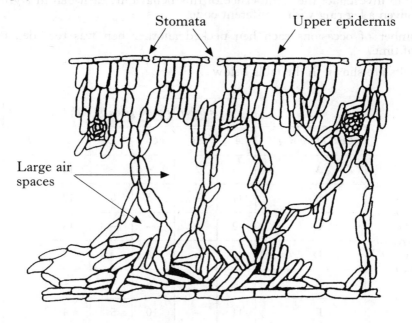

Explain how each of the following features is an adaptation which is characteristic of a hydrophyte with floating leaves.

1. Stomata present only on the upper leaf surface.

_____

_____

_____ **(1)**

2. Presence of large air spaces.

_____

_____

_____ **(1)**

Marks

11. Hens living in groups frequently peck each other.

In order to investigate the significance of this behaviour, each hen in a group of six was given a leg ring with a different code.

The number of occasions each hen pecked another hen was recorded over a period of time.

The results are shown in the table below.

	Code of leg ring	A	B	C	D	E	F
		Number of pecks given by each bird					
	A	–	2	–	–	10	–
	B	17	–	5	6	12	9
Number of pecks received by each bird	C	2	–	–	–	13	–
	D	6	–	8	–	7	–
	E	–	–	–	–	–	–
	F	11	–	10	5	4	–

(a) From the results, give the pecking order of the hens starting from the most dominant hen.

□ → □ → □ → □ → □ → □

(2)

(b) (i) Many mammals which form social groups demonstrate dominance hierarchies.

What is meant by the term "dominance hierarchy"?

_____

_____

_____ **(1)**

(ii) Give **one** advantage of a dominance hierarchy in a social organisation.

_____

_____

_____ **(1)**

Marks

12. Lions can be solitary hunters or may form cooperative hunting groups. The graph below shows how the number of lions in a hunting group affects hunting success and food intake per lion when hunting the same type of prey.

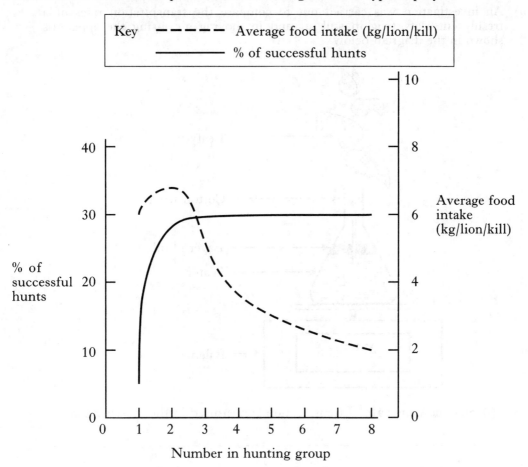

(*a*) Identify the group size which provides the greatest average food intake per lion.

Group size _____ **(1)**

(*b*) Which **two** of the factors listed below would give an advantage to lions which hunt in groups rather than as individuals?

Tick (✓) **two** boxes.

List

1. A group of hyenas may displace a solitary lion from its kill. ☐

2. Prey populations are abundant. ☐

3. Available prey are large animals. ☐ **(1)**

(*c*) From the data, explain why the average intake of food per lion in a group of 8 is less than that in a group of 4.

_____

_____ **(2)**

SECTION B

Answer ALL questions in this section.

Marks

13. (a) An investigation was carried out to compare the transpiration rates of a freshly cut leafy twig in still air and in moving air, using the apparatus shown in the diagram below.

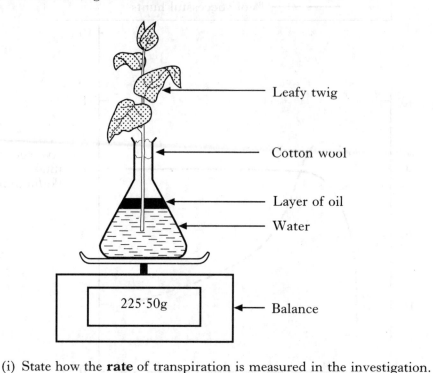

(i) State how the **rate** of transpiration is measured in the investigation.

_____

_____

_____ **(2)**

(ii) Explain why the results would be unreliable if the layer of oil was not included.

_____

_____

_____ **(1)**

(iii) Identify **two** environmental factors which had to be kept constant to allow a fair comparison between the results obtained in still and in moving air.

Factor 1. _____ **(1)**

Factor 2. _____ **(1)**

(iv) Give **one** way in which the reliability of the results could have been improved.

_____ **(1)**

Marks

(v) The apparatus was left in the conditions of the room for 30 minutes before starting to measure the transpiration rate.
Explain why this was good experimental procedure.

_____

_____ **(1)**

(vi) If a balance was not available, describe one other method which could be used to measure transpiration rate.

_____

_____ **(1)**

(b) The results in the table below show the mass of water lost from a leafy twig over a five hour period in still air.

Time (hours)	0	1	2	3	4	5
Total mass of water loss (g)	0	9	18	38	47	56

(i) Using the results, plot a **line graph** of water loss against time on the grid provided below. **(2)**

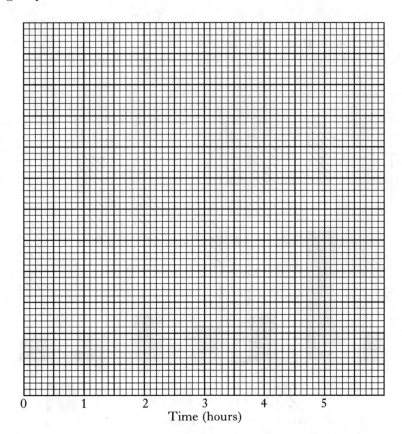

Time (hours)

(ii) What evidence from the results suggest that environmental conditions were not constant during the experiment?

_____

_____ **(1)**

Marks

14. The table below shows the average yield in tonnes per hectare, for the years 1890 and 1990 for four Scottish crops.

Crop	Average yield in tonnes per hectare	
	1890	1990
Potato	15·0	37·5
Turnip	4·5	10·0
Barley	2·4	6·6
Wheat	2·7	8·1

(*a*) Which Scottish crop has shown the greatest percentage increase in average yield per hectare for the period recorded?

Space for calculation

Crop _____ **(1)**

The bar chart below shows the annual percentage loss of yield of crops world-wide in 1990 as a result of the effects of weeds, disease and insects.

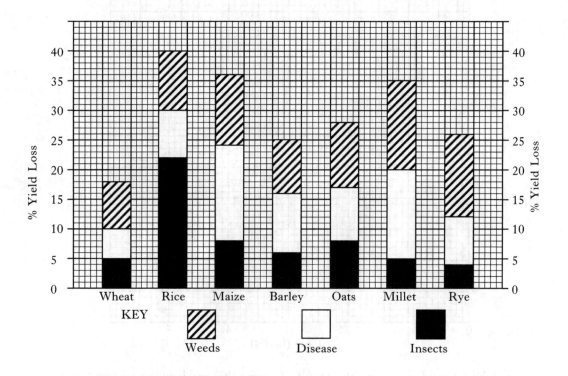

Marks

(b) Complete the table below to show which world crop plant has the lowest percentage annual loss of yield due to each of the factors listed.

Factor	Crop Plant
Presence of weeds	
Disease	
Infestation by insects	

(2)

(c) Complete the table below to show which world crop plant would have the greatest percentage increase in yield as a result of the following treatments.

Treatment	Crop plant showing the greatest increase in yield
Application of insecticide	
Somatic fusion technique to produce plants resistant to disease	
Application of selective weed-killer	

(2)

(d) Suggest one factor other than the control of weeds, disease and insects which would account for the increase in average yield of crops between 1890 and 1990.

_____

(1)

(e) In 1990, a Scottish farmer growing 10 hectares of wheat lost 5% of the crop during storage as a result of mice.
Calculate his final yield in tonnes.
Space for calculation

Answer _____ tonnes   **(1)**

(f) Express, as the smallest whole number ratio, the percentage loss in the maize crop caused by each of the factors.
Space for calculation

_____	:	_____	:	_____	
Insects		Disease		Weeds	**(1)**

Marks

15. A biotechnology company carried out an investigation into the production of a protease enzyme by a bacterium.

The bacterium was grown in a fermenter with glucose as the food source.

The bacterium releases the protease enzyme into the growth medium.

Graph 1 shows changes in the glucose concentration, the concentration of protease and the dry mass of bacteria during 60 hours growth.

Graph 1

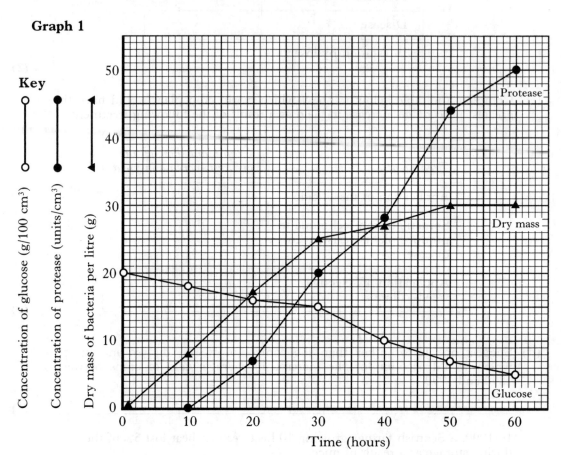

(a) From **Graph 1**, calculate the average increase per hour in the dry mass of bacteria during the investigation.

Space for calculation

Answer _____ g/hour **(1)**

(b) The company uses the following formula to measure bacterial yield.

$$\text{Bacterial yield} = \frac{\text{Increase in dry mass of bacteria (g/litre)}}{\text{Glucose consumed (g/100 cm}^3)}$$

Calculate the bacterial yield after 30 hours using the formula.

Space for calculation

Yield _____ **(1)**

Marks

(c) What is glucose used for

1. between 0 and 10 hours?

_____

_____ **(1)**

2. between 50 and 60 hours?

_____

_____ **(1)**

(d) From **Graph 1**, during which 10 hour period was protease production greatest?

Tick the correct box.

0–10 hours ☐   10–20 hours ☐   20–30 hours ☐

30–40 hours ☐   40–50 hours ☐   50–60 hours ☐   **(1)**

(e) How many hours passed before 50% of the glucose was used up?

_____ **(1)**

Graph **2** shows the mass of bacteria after 60 hours growth when the investigation was repeated at different oxygen concentrations.

Graph **3** shows the protease concentration after 60 hours growth at different oxygen concentrations when magnesium was present and when magnesium was absent from the growth medium.

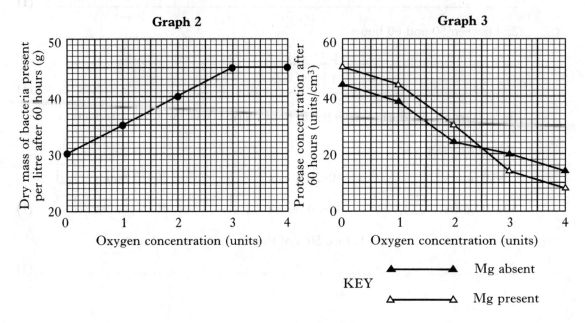

Graph 2

Graph 3

KEY

Mg absent

Mg present

(*f*) From **Graph 2**, identify the range of oxygen concentrations which has a limiting effect on the growth of bacteria.

_____ **(1)**

(*g*) From **Graph 3**, describe the effect of the presence of magnesium on protease production at low and at high oxygen concentrations.

Low oxygen concentration _____

High oxygen concentration _____ **(1)**

(*h*) From **Graphs 2 and 3**, identify the conditions which were present to give the results shown in **Graph 1**.

Oxygen concentration _____ units **(1)**

Magnesium present or absent _____ **(1)**

SECTION C

Both questions in this section should be attempted.
Note that each question contains a choice.

Questions 16 and 17 should be attempted on the blank pages which follow.

Supplementary sheets, if required, may be obtained from the invigilator.

Labelled diagrams may be used where appropriate.

Marks

16. Answer **either** A **or** B.

    A.   Give an account of photosynthesis under the following headings:

       (*a*)   Extraction and separation of the photosynthetic pigments by means of paper chromatography;   **6**

       (*b*)   Structure of a chloroplast;   **3**

       (*c*)   The light stage of photosynthesis.   **6**

                      (15)

OR

    B.   Give an account of homeostasis in animals under the following headings:

       (*a*)   The principle of negative feedback;   **2**

       (*b*)   The role of the pituitary gland and ADH in water balance;   **6**

       (*c*)   The role of hormones in the regulation of blood sugar concentration.   **7**

                      (15)

17. Answer **either** A **or** B.

    A.   Give an account of natural selection and the evolution of new species.   **(15)**

OR

    B.   Give an account of the effects of IAA and GA in the control of growth and development in plants.   **(15)**

[END OF QUESTION PAPER]

SCOTTISH
CERTIFICATE OF
EDUCATION
1996

MONDAY, 13 MAY
9.30 AM – 12.00 NOON

BIOLOGY
HIGHER GRADE
Paper II

INSTRUCTIONS TO CANDIDATES

1. All questions should be attempted: it should be noted however that questions in Section C each contain a choice.

2. The questions may be answered in any order but all answers are to be written in the spaces provided in this answer book, and must be written clearly and legibly in ink.

3. Additional space for answers and rough work will be found at the end of the book. If further space is required, supplementary sheets may be obtained from the Invigilator and should be inserted inside the front cover of this booklet.

4. The number of questions must be clearly inserted with any answers written in the additional space.

5. Rough work, if any should be necessary, should be written in this booklet and then scored through when the fair copy has been written.

6. Before leaving the examination room this book must be given to the Invigilator. If you do not, you may lose all the marks for this paper.

SECTION A

Answer ALL questions in this section.

Marks

1. (*a*) The diagrams below represent some of the structures present in animal and plant cells.

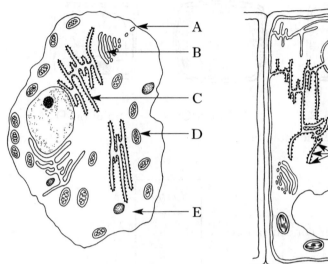
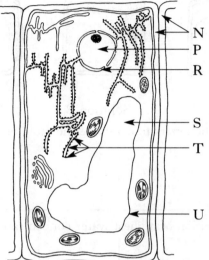

Use letters from the diagrams to match the functions in the table below.
(Each letter can be used **once**, **more than once** or **not at all**.)

Function	Letter
Site of m-RNA synthesis	
Site of anaerobic respiration	
Site of protein synthesis	
Transport of protein	
Controls movement of materials into and out of the cell	
Packaging of secretions	

(3)

(*b*) The same plant cell was placed in four solutions, each with a different *Marks* water concentration.

The drawings below show its appearance after several minutes in each solution.

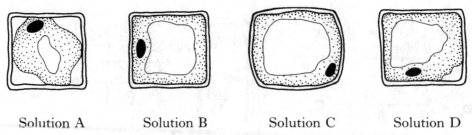

Solution A   Solution B   Solution C   Solution D

Use the letters to complete the table below.

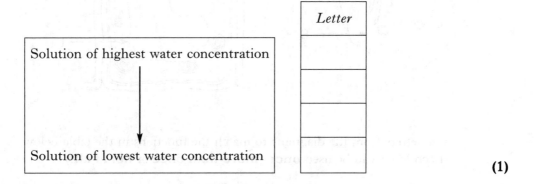

	Letter
Solution of highest water concentration	
↓	
Solution of lowest water concentration	

(1)

2. (*a*) The following graph shows the absorption spectrum of a photosynthetic pigment and the rate of photosynthesis by a green plant, over the same range of colours of light.

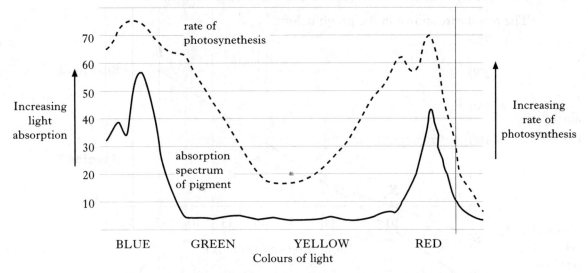

(i) Name a photosynthetic pigment which shows this absorption spectrum.

Pigment _____

(1)

(ii) Not all of the light energy which strikes a leaf is absorbed.

State **two** possible fates of the light energy which is not absorbed.

1. _____

2. _____

(1)

(iii) Other pigments are involved in photosynthesis.

Explain how the data in the graph support this statement.

_____

_____

(1)

(*b*) The following processes occur in either the light-dependent stage or the Calvin cycle.

Place ticks (✓) in the appropriate boxes to indicate the **two** processes which occur in the light-dependent stage.

Processes	Boxes
Formation of GP (PGA)	
Splitting of water molecules	
Generation of ATP	
Release of hydrogen from the reduced hydrogen acceptor (NADPH$_2$)	

(1)

31

(c) A shade plant and a sun plant had the rates of carbon dioxide *Marks* exchange by their leaves measured at different light intensities.

The results are shown in the graph below.

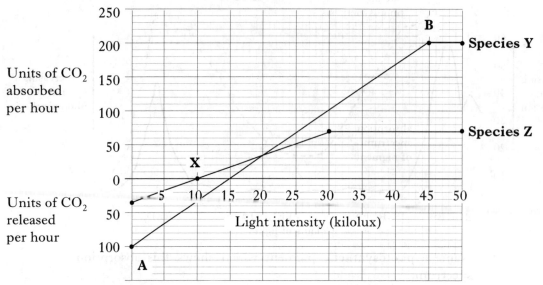

(i) Point **X** on the graph indicates the compensation point for **species Z**. By how many kilolux does the value of this light intensity have to be increased before **species Y** is at compensation point?

Space for calculation

Answer _____ kilolux **(1)**

(ii) Which species is a shade plant? Explain your choice of answer.

Species _____

Explanation _____

_____ **(1)**

(iii) From the data, name the environmental factor which limits the rate of photosynthesis for **species Y** between points **A** and **B**.

Factor _____ **(1)**

(iv) When the rate of photosynthesis is the same for both species of plant,

1. the value of the light intensity is _____ kilolux

2. the number of units of carbon dioxide absorbed per hour is _____ units. **(2)**

Marks

3. The diagram below represents an outline of four stages of respiration in muscle cells.

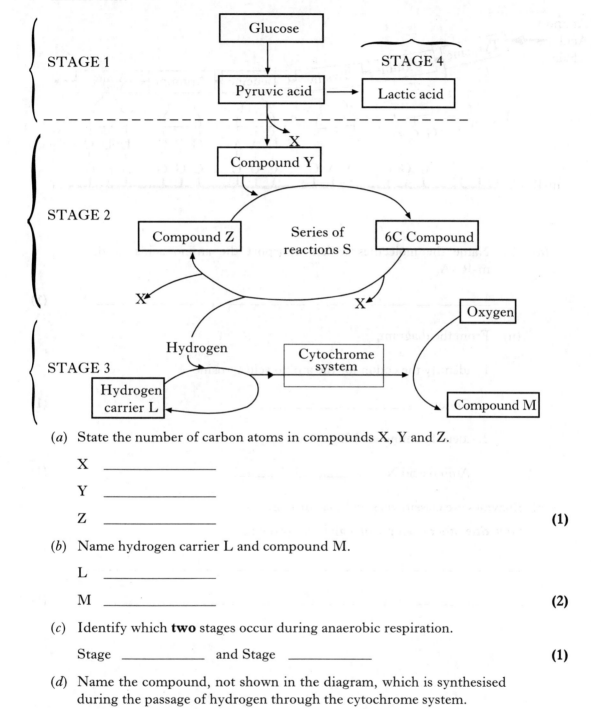

(a) State the number of carbon atoms in compounds X, Y and Z.

X _____

Y _____

Z _____ (1)

(b) Name hydrogen carrier L and compound M.

L _____

M _____ (2)

(c) Identify which **two** stages occur during anaerobic respiration.

Stage _____ and Stage _____ (1)

(d) Name the compound, not shown in the diagram, which is synthesised during the passage of hydrogen through the cytochrome system.

Name _____ (1)

(e) Name the series of reactions S and state their exact location within the mitochondrion.

Name _____

Location _____ (2)

4. The diagram below represents a stage in protein synthesis. *Marks*

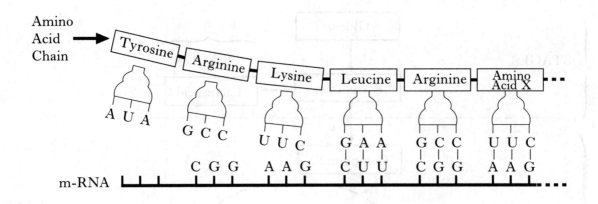

(*a*) (i) Name the molecules which transport the amino acids to the m-RNA.

_____ **(1)**

(ii) From the diagram,

1. identify the codon for the amino acid tyrosine

   Codon _____ **(1)**

2. identify amino acid X.

   Amino acid X _____ **(1)**

(*b*) Enzymes are classified as globular protein.

Give **one** other example of a globular protein.

_____

_____ **(1)**

5. (*a*) The coat colour in mice is controlled by two genes located on different   *Marks*
chromosomes.

Each gene has two alleles. **A** is dominant to **a**, and **B** is dominant to **b**.
The presence of the allele **A** always produces **grey** coat colour.

The other possible genotypes and their phenotypes are shown in the
table below.

Genotype	Phenotype
aaBB	Black coat
aaBb	Black coat
aabb	Brown coat

Male mice heterozygous for both genes were crossed with female
brown-coated mice.

(i) Complete the table below for this cross.

	Male	*Female*
Parental phenotype		Brown coat
Parental genotype		
Genotype of gametes		

(1)

(1)

(1)

(ii) What are the possible F$_1$ genotypes?
Space for working

F$_1$ genotypes: _____ _____ _____ _____   (1)

(iii) State the expected F$_1$ phenotype ratio.
Space for calculation

Ratio _____ Grey: _____ Black: _____ Brown   (1)

(*b*) In *Drosophila*, the genes for wing length (W), eye colour (E), body colour (B) and presence of bristles (P) are linked.

The table below gives the frequency of recombination obtained in crosses involving different pairs of linked genes.

Gene pair in cross			Frequency of recombination
Wing length	×	Eye colour	12 %
Wing length	×	Body colour	18 %
Wing length	×	Presence of bristles	15 %
Eye colour	×	Body colour	6 %
Body colour	×	Presence of bristles	3 %

Use the information to show the position of these genes in relation to each other on the chromosome diagram below.

Use the letters W, E, B and P to identify the position of each gene on the chromosome.

Chromosome ———|——————————————|——|——|—— **(1)**

(*c*) The diagram below shows a cell from a flowering plant.

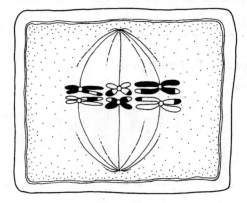

(i) Which **two** features in the diagram show that the cell is dividing by meiosis and not by mitosis?

1. _____

2. _____ **(2)**

(ii) Name a structure in a flower where this cell may be located.

Structure _____ **(1)**

6. The Galapagos islands and seed-eating finches which live there are described in the following statements.

1. The Galapagos islands are of volcanic origin.

2. The islands are 600 miles from the mainland of South America.

3. Many species of plants and animals which are found nowhere else in the world inhabit the islands.

4. Among the animals there is a wide variety of finch species.

5. Most biologists agree that these species evolved from the one common ancestral species of seed-eating finch.

6. These finch species now occupy many of the niches in the ecosystems of the islands.

7. Similar niches on the mainland are filled by entirely different species of birds.

(*a*) Use the numbers to identify the **two** statements which suggest that the evolution of these finches is an example of adaptive radiation.

Statement number _____ and _____ **(2)**

(*b*) Two of the species of seed-eating finch found on the islands are shown in the diagrams below (drawn to same scale).

Species A   **Species B**

Describe an adaptation of **species A**, shown in the diagram, which allows this species to avoid competing for food with **species B**.

Explain how this adaptation allows **species A** to avoid competition for food with **species B**.

Adaptation _____

_____ **(1)**

Explanation _____

_____ **(1)**

(*c*) The diagrams below show the preparation of a cell for somatic fusion.

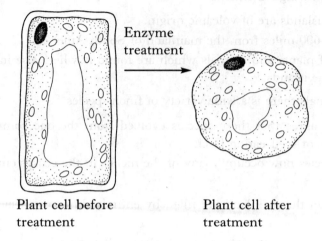

Enzyme
treatment

Plant cell before
treatment

Plant cell after
treatment

(i) Name the enzyme used in the treatment above.

_____ **(1)**

(ii) What name is given to the cell after treatment?

_____ **(1)**

(iii) What problem in plant breeding has been overcome by the use of somatic fusion?

_____

_____ **(1)**

Marks

7. (*a*) The diagrams show the results of an experiment in which apical buds were removed from plants and the decapitated shoots treated as indicated.

Treatment A   Treatment B   Treatment C

Paste with plant growth substance X — lateral bud developing

Paste with plant growth substance Y

Paste with no plant growth substance — lateral bud developing

(i) Which two treatments, when compared, demonstrate the role of a growth substance in apical dominance? Explain your choice.

Treatments _____ and _____   **(1)**

Explanation _____

_____   **(1)**

(ii) Write the name of the plant growth substance which is responsible for apical dominance.

Name _____   **(1)**

(*b*) The aleurone layers were removed from ungerminated barley grains and immersed in a solution of gibberellic acid.

After 48 hours, the enzyme amylase was found to be present in the solution. This procedure was repeated using a solution of gibberellic acid and actinomycin. Actinomycin blocks mRNA synthesis.

(i) Predict what effect the presence of actinomycin would have on amylase production. Explain your answer.

Effect on amylase production _____

Explanation _____

_____   **(1)**

(ii) Name the part of a barley grain which produces gibberellic acid.

Name _____   **(1)**

8. (a) The diagram below shows the metabolic pathways by which *Marks* phenylalanine may be metabolised in the human body.

METABOLIC PATHWAY 1

Phenylalanine ⟶ **Tyrosine**

     METABOLIC
     PATHWAY 2

Phenylketone

   (i) Describe the role of genes in metabolic pathways.

      _____

      _____ **(1)**

  (ii) Metabolic pathway 2 operates only when metabolic pathway 1 is blocked. Explain how such a blockage might occur.

      _____

      _____

      _____ **(2)**

(b) The bacterium *Escherichia coli* produces a lactose-digesting enzyme only in the presence of lactose.

The production of this enzyme illustrates the Jacob-Monod hypothesis of gene control in bacteria.

The table below contains statements which refer to events in this control process.

Write **TRUE** or **FALSE** in the table to indicate which events occur in the **absence** of lactose.

List of events in gene control process	Occurs in the absence of lactose— TRUE or FALSE
Repressor molecule combines with operator	
Structural gene "switched on"	
Repressor-inducer complex formed	
Regulator gene produces repressor molecule	

                                                   (2)

9. (*a*) Enzymes in the liver break down alcohol. *Marks*

The graph below shows the effect of different concentrations of lead on the breakdown of alcohol by these enzymes.

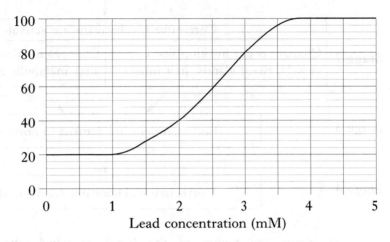

Alcohol concentration after 30 minutes as a % of initial concentration

Lead concentration (mM)

Complete the table below by describing the effect of each of the following lead concentrations on the activity of the enzymes.

Lead concentration (mM)	Effect on enzyme activity
0·5	
2·5	
4·5	

(2)

(*b*) State **one** possible effect of each of the following on human fetal development.

1. Excessive alcohol consumption during pregnancy

_____

_____ (1)

2. Use of the drug thalidomide during pregnancy

_____

_____ (1)

10. (*a*) The diagram below shows stages in the control of body temperature in a mammal.

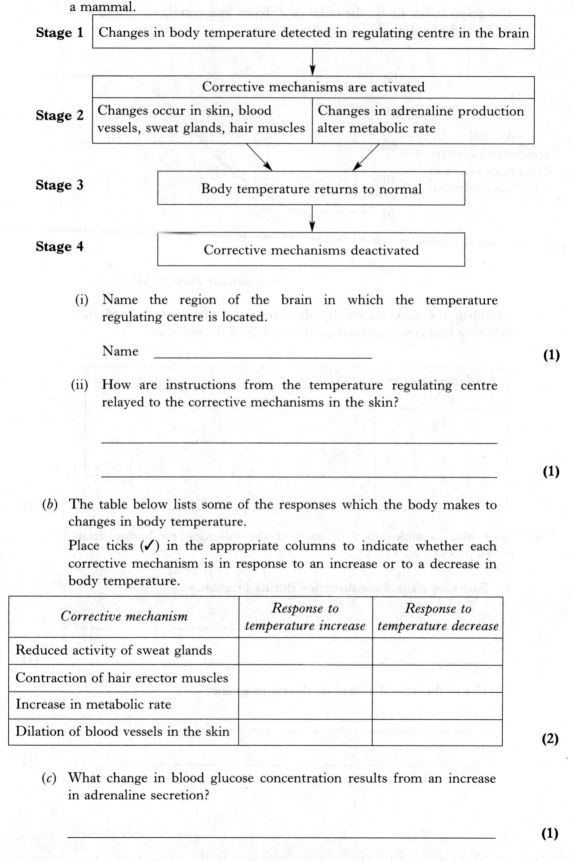

Stage 1 | Changes in body temperature detected in regulating centre in the brain

Corrective mechanisms are activated

Stage 2 | Changes occur in skin, blood vessels, sweat glands, hair muscles | Changes in adrenaline production alter metabolic rate

Stage 3 | Body temperature returns to normal

Stage 4 | Corrective mechanisms deactivated

   (i) Name the region of the brain in which the temperature regulating centre is located.

   Name _____ **(1)**

   (ii) How are instructions from the temperature regulating centre relayed to the corrective mechanisms in the skin?

   _____

   _____ **(1)**

(*b*) The table below lists some of the responses which the body makes to changes in body temperature.

Place ticks (✓) in the appropriate columns to indicate whether each corrective mechanism is in response to an increase or to a decrease in body temperature.

Corrective mechanism	Response to temperature increase	Response to temperature decrease
Reduced activity of sweat glands		
Contraction of hair erector muscles		
Increase in metabolic rate		
Dilation of blood vessels in the skin		

(2)

(*c*) What change in blood glucose concentration results from an increase in adrenaline secretion?

_____ **(1)**

11. (*a*) The diagrams below relate to movement of water in the transpiration *Marks*
stream. Table 1 describes the movement of water at different
locations within the plant.

Table 1

Location	Movement of water
A	Uptake by root
B	Passage through cortex
C	Ascent in xylem
D	Passage through spongy mesophyll
E	Loss through stomata

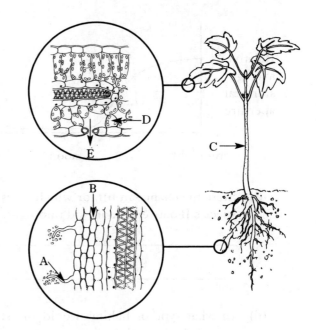

Table 2 below contains explanations which relate to the movement of
water in some of these locations.

Complete the table.

Table 2

Location(s)	Explanation of water movement
	Columns of water maintained by adhesion and cohesion
	Cells gain water from adjacent hypotonic cells
A	

(3)

Marks

(b) The graph below shows changes in the stomatal aperture of two plant species during 24 hours.

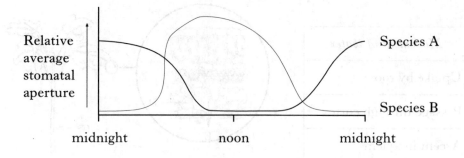

Relative average stomatal aperture

Species A

Species B

midnight   noon   midnight

(i) State the change in turgor which takes place in the guard cells of species B between noon and midnight.

_____

_____ **(1)**

(ii) In what type of habitat would the stomatal rhythm shown by species A be an advantage?

_____

_____ **(1)**

(iii) The transpiration stream supplies water to a plant for support and chemical reactions.
State **one** other benefit which the plant obtains from the transpiration stream.

_____

_____ **(1)**

SECTION B

Answer ALL questions in this section.

Marks

12. The diagrams below show stages in the procedure used to extract and identify the photosynthetic pigments present in fresh leaves.

Stage 1 ⟶ **Stage 2** ⟶ **Stage 3** ⟶ **Stage 4**

Extract pigments   Separate leaf debris   Load pigments on to   Develop the
                from solution of pigments   chromatography paper   chromatogram

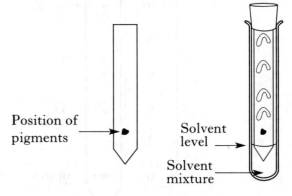

Position of pigments

Solvent level

Solvent mixture

(a) Describe how the pigments were extracted from the leaf cells in **Stage 1**.

_____

_____

_____ **(1)**

(b) Name a technique used to separate the leaf debris from the solution of pigments at **Stage 2**.

Technique   _____ **(1)**

(c) Name the **two** components of the solvent mixture used in **Stage 4**.

Components 1. _____ 2. _____ **(1)**

(d) Explain the importance of the position of the pigments on the chromatography paper as shown in the **Stage 3** diagram.

_____

_____ **(1)**

(*e*) During **Stage 4**, the solvent rose to the line indicated by X in the diagram below.

The position of each pigment was measured using a ruler as shown in the diagram.

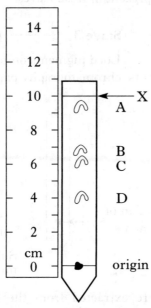

(i) Complete the table below to show the names, the colours and the maximum distances travelled by each pigment.

Letter	Name of pigment	Colour	Maximum Distance travelled (cm)
A		Yellow	9·8 cm
B	Xanthophyll		
C			6·5 cm
D		Green	4·4 cm

(3)

(ii) Occasionally results are obtained in which the pigments are all carried to the top of the chromatography paper.

What error in experimental procedure could account for these results?

_____

_____ **(1)**

13. *Dendroica* are birds which nest and feed in fir trees.

Investigators recorded the feeding patterns of three different species of *Dendroica* on tree diagrams as shown below.

Tree diagram

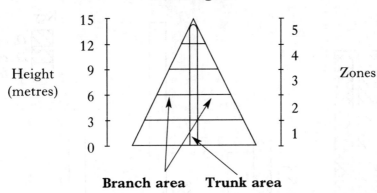

Branch area   Trunk area

Shading in the following tree diagrams shows how often each of the species feeds within different parts of the tree.

Key

Feeds often
Feeds occasionally
Feeds rarely
Never feeds

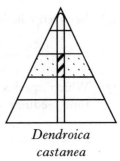

Dendroica coronata   *Dendroica tigrina*   *Dendroica castanea*

(a) Which **one** of the species spends most of its feeding time within the trunk area of the tree?

Species _____ **(1)**

(b) Identify the **two** species which feed between a height of 9 and 12 metres.

Species 1. _____ 2. _____ **(1)**

(c) With reference to the feeding zones, describe fully the feeding pattern of *Dendroica tigrina*.

_____

_____

_____ **(2)**

(d) What evidence is there that the three species are able to avoid interspecific competition for food?

_____ **(1)**

Marks

The percentage of nests of each species present within each zone is shown in the bar chart below.

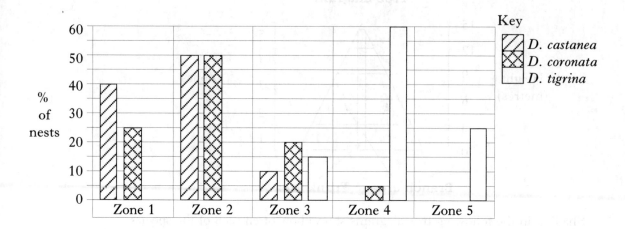

(e) Between which two species is there the least competition for nest sites?

_____ **(1)**

(f) Which species nests most often in the zone in which it spends most time feeding?

Species _____ **(1)**

(g) In the table below, when each of the statements is matched to each of the three species, only six of the combinations are true.

Identify with the letter **T** in the appropriate boxes in the table the six combinations which are true.

Statement	D. coronata	D. tigrina	D. castanea
Spends most of its feeding time in the outer branches			
Spends time feeding within more than one zone			
May spend some time on the ground when feeding			
Nests in four different zones			

(2)

14. Glucose is stored in muscle tissue in the form of glycogen. Glycogen is the main energy source for respiring muscle tissue.

Marks

Endurance time is the time that exercise can be maintained until the muscles become exhausted.

Graph A shows the average endurance time for groups of volunteers with different initial concentrations of glycogen in their muscles.

Graph A

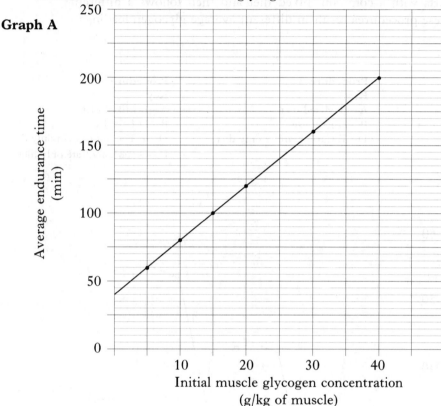

Initial muscle glycogen concentration
(g/kg of muscle)

(a) From **Graph A**, calculate the difference in the average endurance time between the group of volunteers who started with an initial muscle glycogen concentration of 5g/kg as against the group of volunteers with an initial muscle glycogen concentration of 30g/kg.

Space for calculation

Difference _____ minutes **(1)**

(b) What evidence in **Graph A** shows that glycogen is not the only source of energy available to the muscles?

_____

_____ **(1)**

(c) From **Graph A**, predict the initial muscle glycogen concentration of an individual with an endurance time of 240 minutes.

Answer _____ g/kg **(1)**

Glycogen loading refers to methods used by athletes to increase the concentrations of glycogen stored in their muscles.

Graph B shows the changes in average muscle glycogen concentrations for three groups of volunteers over periods of time.

Each group starts with a common mixed diet and then follows a different procedure. These procedures result in different average glycogen loadings.

Procedure 1
Key
A—B Mixed diet
B—C Carbohydrate only diet

Procedure 2
Key
A—B Mixed diet
B—D Exercise
D—E Carbohydrate only diet

Procedure 3
Key
A—B Mixed diet
B—D Exercise
D—F Fat and protein only diet
F—G Carbohydrate only diet

Graph B

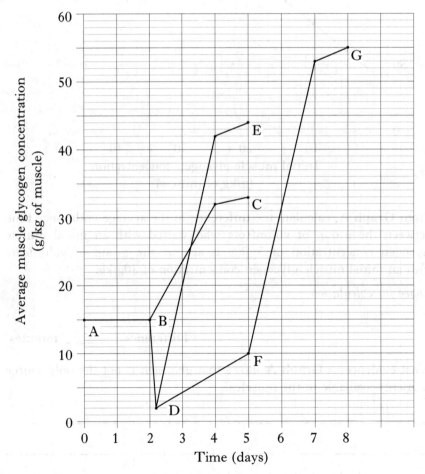

(*d*) Identify **two** ways in which the average glycogen loading achieved by a carbohydrate only diet can be improved.

1. _____

2. _____ **(2)**

(*e*) Express, as the **simplest whole number ratio**, the average muscle glycogen concentrations achieved at the end of the three procedures.

Space for calculation

Procedure 1 : Procedure 2 : Procedure 3

Ratio _____ : _____ : _____ **(1)**

(*f*) Calculate the percentage by which the final average muscle glycogen concentration obtained by procedure 3 exceeds that obtained by procedure 2.

Space for calculation

Percentage increase _____ **(1)**

(*g*) From **Graphs A** and **B**, calculate the expected endurance time of a volunteer who had followed procedure 1.

Space for calculation

Endurance time _____ minutes **(1)**

SECTION C

Both questions in this section should be attempted.
Note that each question contains a choice.

Questions 15 and 16 should be attempted on the blank pages which follow.

Supplementary sheets, if required, may be obtained from the invigilator.

Labelled diagrams may be used where appropriate.

Marks

15. Answer **either** A **or** B.

    A. Give an account of DNA under the following headings:

        (*a*) DNA replication;   6

        (*b*) genetic engineering;   4

        (*c*) gene mutations.   5

                                                                                  (15)

    OR

    B. Give an account of growth and development in plants under the following headings:

        (*a*) position and activity of meristems;   3

        (*b*) formation of xylem vessels and sieve tubes from undifferentiated cells;   6

        (*c*) importance of magnesium and phosphorus.   6

                                                                                  (15)

16. Answer **either** A **or** B.

    A. Give an account of how water balance is maintained in fish and in desert mammals.   (15)

    OR

    B. Give an account of the principle of negative feedback as illustrated by control of water content and glucose concentration of blood in a mammal.   (15)

[*END OF QUESTION PAPER*]

SCOTTISH
CERTIFICATE OF
EDUCATION
1997

MONDAY, 12 MAY
9.30 AM – 12.00 NOON

BIOLOGY
HIGHER GRADE
Paper II

INSTRUCTIONS TO CANDIDATES

1. All questions should be attempted: it should be noted however that questions in Section C each contain a choice.

2. The questions may be answered in any order but all answers are to be written in the spaces provided in this answer book, and must be written clearly and legibly in ink.

3. Additional space for answers and rough work will be found at the end of the book. If further space is required, supplementary sheets may be obtained from the Invigilator and should be inserted inside the front cover of this booklet.

4. The number of questions must be clearly inserted with any answers written in the additional space.

5. Rough work, if any should be necessary, should be written in this booklet and then scored through when the fair copy has been written.

6. Before leaving the examination room this book must be given to the Invigilator. If you do not, you may lose all the marks for this paper.

SECTION A

Answer ALL questions in this section.

1. The following numbered statements refer to respiration.

    1. In respiration, a high energy compound is synthesised which drives the energy-requiring processes in cells.

    2. In the first stage of respiration, a 6-carbon sugar is broken down to two molecules of a 3-carbon compound.

    3. If oxygen is present, the 3-carbon compound is converted to a 2-carbon compound.

    4. This 2-carbon compound enters a cycle of reactions, during which carbon dioxide and hydrogen are released.

    5. The hydrogen released is transferred to the cytochrome system and it is at this stage that oxygen plays its role in respiration.

    6. In the absence of oxygen, the less efficient process of anaerobic respiration takes place and different metabolic products are formed.

Marks

(a) Name the high energy compound referred to in statement 1.

   Name of compound _____   **(1)**

(b) Give **one** example of the energy-requiring processes referred to in statement 1.

   Example _____

   _____   **(1)**

(c) Which statement describes glycolysis?

   Number of statement _____   **(1)**

(d) What is the exact location of the reactions mentioned in statement 4?

   _____   **(1)**

(e) Name the compound responsible for hydrogen transfer in statement 5.

   Name of compound _____   **(1)**

(f) Describe the role of oxygen referred to in statement 5.

   _____   **(1)**

(g) Explain why the term "less efficient" (statement 6) is used when comparing anaerobic respiration to aerobic respiration.

   _____

   _____   **(1)**

(h) Name the final metabolic product of anaerobic respiration in human muscle cells.

   Name _____   **(1)**

Marks

2. The stages shown below take place when a cell is invaded by a virus.

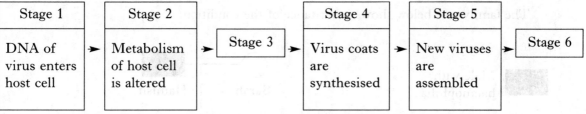

(a) Describe the processes which take place during Stages 3 and 6.

Stage 3 _____

_____

_____

Stage 6 _____

_____

_____ **(2)**

(b) Enzymes of the host cell are used during Stage 4.

Name **one** other substance which is supplied by the host cell during this stage.

Name of substance _____ **(1)**

(c) Why are ribosomes required for Stage 4?

_____

_____

_____ **(1)**

Marks

3. In humans, the allele for haemophilia (h) is sex linked and recessive to the normal allele (H).

The family tree below shows inheritance of the condition.

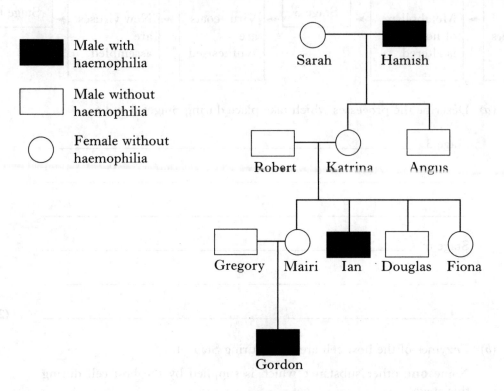

■ Male with haemophilia

□ Male without haemophilia

○ Female without haemophilia

(*a*) Give the genotype of each of the following individuals.

1. Gordon _____

2. Gregory _____

3. Mairi _____ **(2)**

(*b*) Individuals heterozygous for the condition are described as "carriers". Name the **two** individuals in the family tree who are **known** to be carriers.

Name 1. _____ 2. _____ **(2)**

Marks

4. (*a*) You have to decide whether each of the following statements about nucleic acids is **TRUE** or **FALSE** and **tick the appropriate box**.

If you decide the statement is **FALSE**, you should then write the **correct word** in the right hand box to replace the word **underlined in the statement**.

Statements	True	False	Correct word
During the formation of a new DNA molecule, base pairing is followed by bonding between deoxyribose and <u>bases</u>			
Synthesis of m-RNA takes place in the <u>nucleus</u>			
m-RNA consists of many <u>codons</u>, each consisting of a base, ribose and phosphate			

(3)

(*b*) m-RNA codes for the amino acids which bond to form a protein chain.

The diagram below can be used to identify the amino acids which are coded for by some m-RNA codons.

For example, the m-RNA codons with base sequences CAU and CAC both code for the amino acid histidine*.

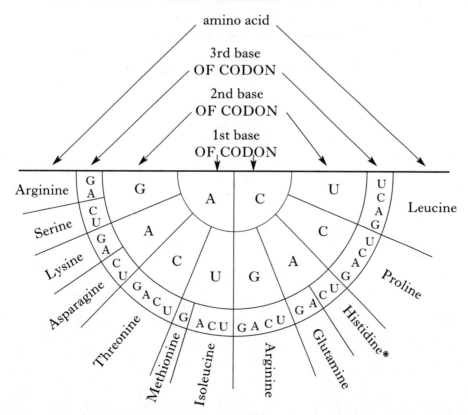

Marks

Use the information in the diagram to answer the following questions.

(i) DNA determines the base sequence on m-RNA.

Which base sequence in DNA would code for the amino acid methionine?

_____   **(1)**

(ii) Which amino acid is carried by the t-RNA molecule which has the anticodon GAC?

_____   **(1)**

(iii) A substitution mutation changes the first base in a codon.

1. Which amino acid will still be inserted correctly into the protein if the first base of one of its codons is changed from A to C?

Amino acid _____   **(1)**

2. Name the **four** amino acids which could still be inserted in the correct sequence despite any substitution in the third base of their codons.

Names 1. _____

2. _____

3. _____

4. _____   **(2)**

Marks

5. (*a*) The diagrams below show **two** possible chromosome mutations which could occur in a chromatid during the process of meiosis.

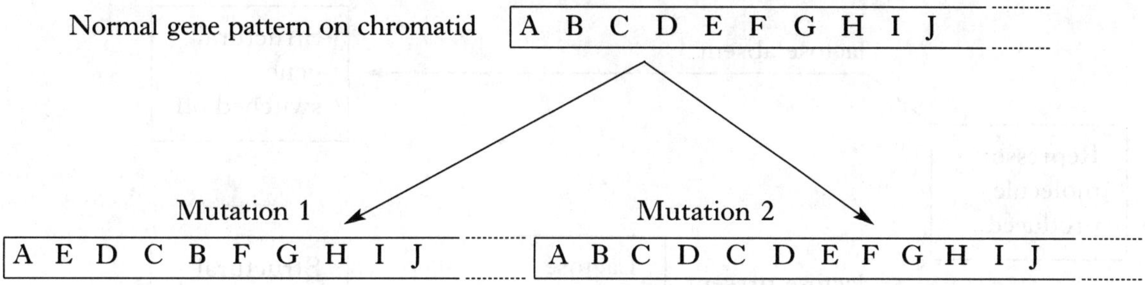

Normal gene pattern on chromatid

| A B C D E F G H I J |

Mutation 1

| A E D C B F G H I J |

Mutation 2

| A B C D C D E F G H I J |

Complete the table below by naming each type of chromosome mutation.

Number	Type of chromosome mutation
1.	
2.	

(2)

(*b*) Name **one** factor, other than chemical agents or temperature changes, which increases mutation rate.

Factor _____

(1)

Marks

6. The diagram below illustrates stages in the control of gene action in bacteria.

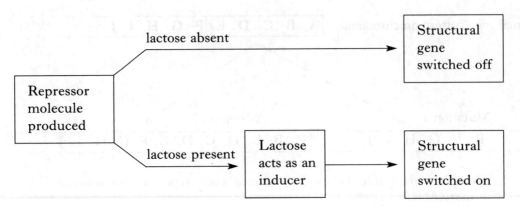

(a) Explain why, if lactose is absent, the structural gene remains switched off.

_____

_____ **(1)**

(b) Explain how lactose acts as an inducer.

_____

_____ **(1)**

Marks

7. The graphs below represent growth patterns in different organisms.

Graph A shows the growth pattern of a locust.

Graph B shows the growth pattern of a tree, such as an oak, during a period of one year.

Graph A

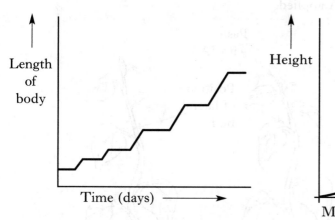

Graph B

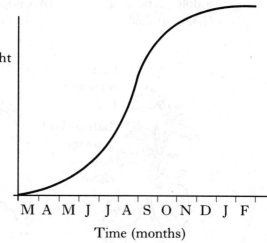

Account for the growth pattern shown by

1. the locust _____

   _____

   _____ **(1)**

2. the tree. _____

   _____

   _____ **(2)**

Marks

8. (*a*) The diagrams below show the results of an investigation in which the apical buds of three plants of the same species were treated as indicated.

Treatment A	Treatment B	Treatment C
Apical bud removed. Paste without indole acetic acid (IAA) applied.	Apical bud removed. Paste containing IAA applied.	Apical bud **NOT** removed.

Paste with IAA

Paste without IAA

Lateral bud growing into a shoot

Position of lateral bud

A   B   C

The apical bud produces IAA. **Explain** how the **results** support this statement.

_____

_____

_____   **(2)**

(*b*) Complete the table below, using the appropriate initial letters, to show whether indole acetic acid **alone** (IAA), gibberellic acid **alone** (GA) or **both** (IAA + GA) have a role in each of the processes listed.

Process	*Initial Letters*
Amylase production by barley grains	
Leaf abscission	
Fruit formation	
Stem elongation	

(2)

Marks

9. (*a*) The diagrams below show four seedlings which have been grown in water culture solutions that differ in the elements they contain.

Experiment A   **Experiment B**   **Experiment C**   **Experiment D**

(i) Identify the elements X and Y.

X _____   Y _____   **(2)**

(ii) Account for the reduced growth in the stem length of the seedling lacking nitrogen.

_____

_____

_____   **(2)**

(iii) If water cultures are not aerated, the uptake of elements is reduced. Account for this fact.

_____

_____

_____   **(2)**

Marks

(b) Many substances affect growth and development in humans.

Insert the appropriate letter from the list of chemical substances into each of the boxes below to match its role in, or effect on, growth and development.

Each letter can be used **once**, **more than once** or **not at all**.

List of Substances   **Boxes**

Name	Letter
Calcium	A
Nicotine	B
Alcohol	C
Vitamin D	D
Lead	E
Thalidomide	F
Magnesium	G
Iron	H

Role in or effect on growth and development	
Inhibits activity of certain enzymes	
Required for the uptake of calcium from the small intestine	
Required for normal growth of teeth and bones	
Required for haemoglobin synthesis	
Retards both physical growth and mental development of the human embryo	

(4)

Marks

10. If a habitat is stripped of its original vegetation, the area is recolonised by plants. The diagram below represents a sequence of plant communities over a period of time in an area where the vegetation had previously been totally destroyed by fire.

Community A (First colonisers)	→	Community B	→	Community C	→	Climax Community D

(*a*) What term is used to describe such a sequence of colonisation?

Term   _____   **(1)**

(*b*) State **two** ways in which the climax community D would be expected to differ from community A.

1. _____

2. _____ **(2)**

(*c*) Give **one** reason to explain why community C could be established only after the area had been colonised previously by community B.

_____

_____ **(1)**

Marks

11. The flow diagram below shows the sequence of events which results from a decrease in the water concentration of the blood.

Events

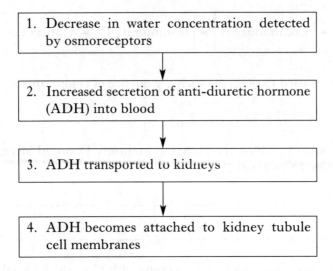

1. Decrease in water concentration detected by osmoreceptors

2. Increased secretion of anti-diuretic hormone (ADH) into blood

3. ADH transported to kidneys

4. ADH becomes attached to kidney tubule cell membranes

(*a*) Name the exact location of the osmoreceptors.

Location _____

(1)

(*b*) The sentences below refer to the changes which occur after stage 4.

Underline **one** of **each pair** of alternatives to make the sentences correct.

The kidney tubule becomes $\begin{Bmatrix} \text{less} \\ \text{more} \end{Bmatrix}$ permeable to water. There is

$\begin{Bmatrix} \text{decreased} \\ \text{increased} \end{Bmatrix}$ reabsorption of water into the bloodstream. The effect of

the change in reabsorption of water is to $\begin{Bmatrix} \text{decrease} \\ \text{increase} \end{Bmatrix}$ the rate at which

urine is produced.

(2)

12. *Ammophila arenaria* (marram grass) grows on sand dunes. Marram grass shows features typical of xerophytes.

Diagram A below shows marram grass growing in a sand dune.

Diagram B below shows a high power view of part of a leaf section of marram grass.

Diagram A

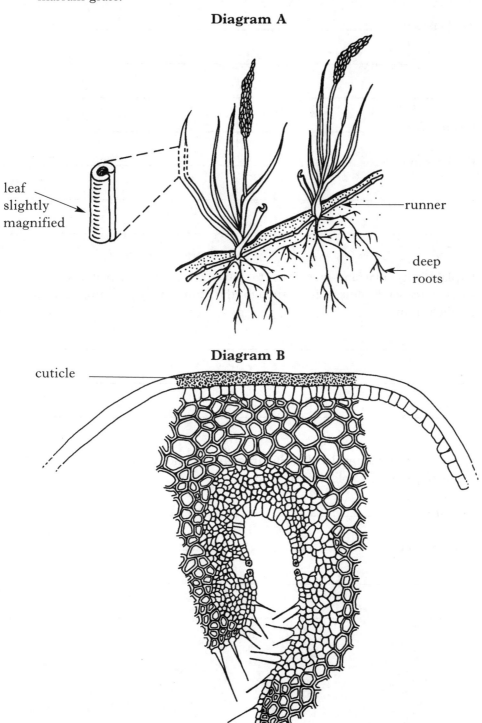

leaf slightly magnified

runner

deep roots

Diagram B

cuticle

Marks

(a) Select **two** features of the leaf shown in the **diagrams** which are typical of a xerophyte. For each feature, explain how water loss is reduced.

Feature 1 _____

Explanation _____

_____

_____ **(1)**

Feature 2 _____

Explanation _____

_____

_____ **(1)**

(b) Explain why marram grass has to have xerophytic adaptations to survive in sand dunes, even though the dunes may be subject to frequent rain.

_____

_____

_____ **(1)**

(c) Explain how a feature labelled in **diagram A** allows marram grass to be an effective first coloniser of sand dunes.

Feature _____

Explanation _____

_____

_____ **(1)**

Marks

13. Bees visit clover flowers to search for nectar and pollen.

Bees were observed feeding from clover flowers within two areas in which the numbers and distribution of flowers were similar.

The distance of the flights made by individual bees when flying from flower to flower was measured.

The results are presented in the pie charts below, and show the percentage of flights made in each distance range within the two areas.

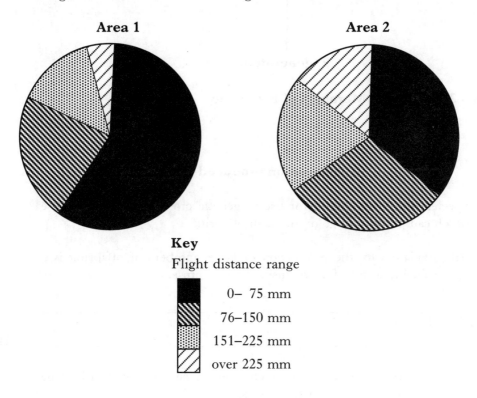

Key

Flight distance range

■	0– 75 mm
	76–150 mm
	151–225 mm
	over 225 mm

(*a*) Explain how the search pattern shown for Area 1 suggests that bees are efficient foragers.

_____

_____   **(2)**

(*b*) In Area 2, there are frequent patches where the flowers have a low nectar and pollen content. As a result, the bees in Area 2 use a different search pattern from those in Area 1.

Explain why the search pattern in Area 2 is of more value to bees when foraging in this area.

Explanation   _____

_____

_____   **(2)**

SECTION B

Answer ALL questions in this section.

Marks

14. Dihybrid crosses can be used to show how the independent assortment of chromosomes during meiosis leads to the production of new phenotypes. The following is a summary of a dihybrid cross to the F_2 generation.

Cross 1   **Parent A**   ×   **Parent B**

F_1 generation

Cross 2   **Members of F_1 crossed**

F_2 generation produced

(a) Species which are chosen for use in genetic crosses are usually those which can produce large numbers of offspring.

   (i) Explain why the production of large numbers of offspring is a valuable feature for organisms used in crosses.

   _____

   _____   **(1)**

   (ii) State **two** other features which a species should have if it is to be chosen for use in genetic crosses.

   Feature 1 _____

   _____

   Feature 2 _____

   _____   **(2)**

(b) Precautions are taken during experimental crosses to ensure that only selected individuals are allowed to interbreed.

   With reference to a named organism, describe **one** such precaution.

   Organism _____

   Precaution _____

   _____   **(1)**

70

The following dihybrid cross was carried out with mice. *Marks*

Parental phenotypes   Black, Straight-haired × Brown, Wavy-haired

F₁ phenotypes   Black, Straight-haired

Members of the **F₁** generation were crossed together

Second cross   Black, Straight-haired × Black, Straight-haired

Results

The table below shows the phenotypes of the **F₂** generation together with their numbers.

Letter	Phenotypes of F₂ offspring	Number of offspring
A	Black, straight-haired	16
B	Black, wavy-haired	6
C	Brown, straight-haired	7
D	Brown, wavy-haired	3

(c)   From the crosses, identify which characteristics are dominant.

_____   **(1)**

(d)   Explain how the results of this cross to the **F₁** generation can show whether the original parents are homozygous or heterozygous.

_____

_____

_____   **(2)**

(e)   Select letters from the table to identify the **two** phenotypes which show that the genes are not linked.

Letters _____ and _____   **(1)**

(f)   The results obtained differ from the theoretically **expected** results.

(i)   From the **total number** of offspring obtained, calculate the number of offspring which would have been expected to have phenotype C.

Space for calculation

Number _____   **(1)**

(ii)   Explain why the results obtained differ from the expected results.

_____

_____   **(1)**

Marks

15. The Pacific sardine was first exploited by fishermen in 1921.

Graph 1 below shows changes in the annual sardine catch along the Pacific coast of North America. (Catch is measured in tonnes.)

Graph 1

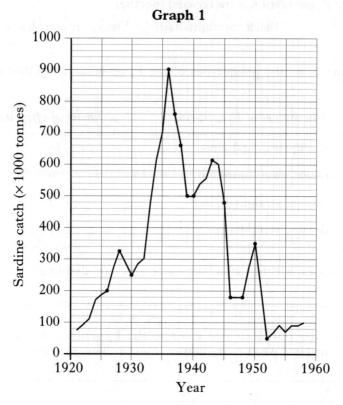

(*a*) From Graph 1, describe the trend in sardine catches between 1930 and 1946.

_____

_____

_____ **(2)**

(*b*) Between which years did the largest **decrease** in catch take place?

Tick the correct box.

1934–35 ☐   1935–36 ☐   1936–37 ☐   1938–39 ☐

1944–45 ☐   1945–46 ☐   1950–51 ☐   1951–52 ☐ **(1)**

(*c*) In how many years was the annual sardine catch greater than 200,000 tonnes?

Space for calculation

Number of years _____ **(1)**

Marks

Graph 2 shows changes in the estimated population of sardines in the same area. (Population is measured in tonnes.)

Graph 2

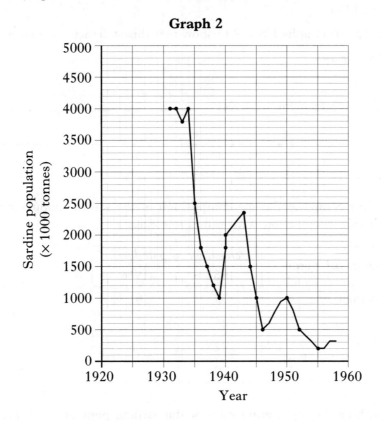

(*d*) From Graph 2, identify the year in which the sardine population first fell to 10% of its maximum.

Space for working

Year _____   **(1)**

(*e*) Describe how the data support the statement that after 1936 the sardine population limited the size of the catch.

_____

_____   **(1)**

Marks

(*f*) When the catch was at its lowest in 1952, the percentage of the sardine population lost due to fishing was less than when the catch was at its highest in 1936.

Use the data in Graphs 1 and 2 to show that this statement is correct.

Space for working

Evidence that statement is correct _____

_____

_____ **(2)**

The table below compares the age composition of the sardine populations in 1935 and 1955. (Population is measured in tonnes.)

Age of fish (years)	Percentage of population	
	1935	1955
less than 3	33	61
3	25	29
4	22	8
over 4	20	2

(*g*) Describe how the age composition of the sardine population in 1955 differs from that of 1935.

_____

_____ **(1)**

(*h*) From Graph 2 and the table, calculate the mass of 4 year old fish present in the sardine population in 1935.

Space for working

Mass _____ × 1000 tonnes **(1)**

Marks

16. For survival of a plant species, it is essential that its seeds must not germinate until environmental conditions are favourable. The seeds of European ash do not germinate until almost two years after they have been dispersed.

The data below show the timescale over which three factors are responsible for preventing germination after seeds have been produced in June.

J F M A M ↑ J A S O N D J F M A M J J A S O N D J F M A

June
(seeds produced)

Bracket shows
germination
period

◄──── year one ────►◄──── year two ────► period

Time (Months)

(*a*) In which two months are impermeability and low temperature reactions the only two factors preventing germination?

Months _____ **(1)**

(*b*) Ash seeds can be made to germinate earlier than they would normally. Suggest a procedure based on the data which could have this effect.

_____

_____

_____ **(1)**

Marks

An investigation was carried out into factors which affect germination in Norway maple seeds.

The table below shows how many seeds germinated after each treatment.

Letter	Treatment of seeds	Number of seeds planted	Number of seeds germinating
A	Seed coat intact. Stored for three months at room temperature.	248	0
B	Seed coat removed. Stored for three months at room temperature.	228	0
C	Seed coat intact. Stored for three months at 2 °C.	235	188
D	Seed coat removed. Stored for three months at 2 °C.	250	190

(c) From the table, identify the factor which encourages germination in Norway maple seeds in this investigation.

Factor _____ **(1)**

(d) Which treatment gave the greatest germination success?

Space for calculation

Letter _____ **(1)**

The graph shows the results of a further investigation into the percentage germination for Norway maple seeds treated as shown in the table below.

	Letter	Treatment
KEY to Graph	E	Chilled for 4 months Then watered with distilled water
	F	Maintained at room temperature for 4 months Then watered with distilled water
	G	Maintained at room temperature for 4 months Then watered with a 5 mg/l solution of abscisic acid
	H	Maintained at room temperature for 4 months Then watered with a 10 mg/l solution of abscisic acid

Marks

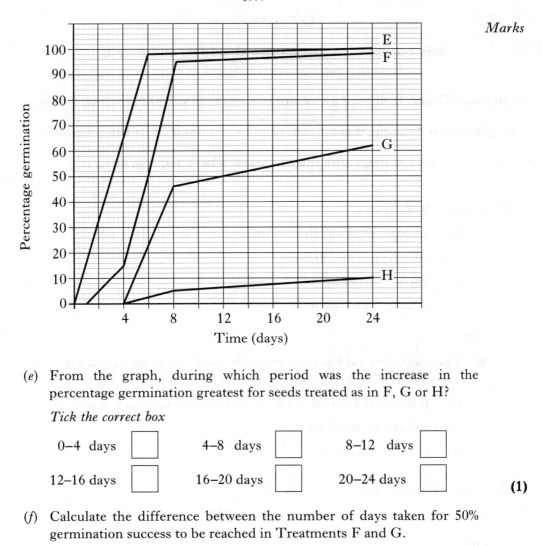

(e) From the graph, during which period was the increase in the percentage germination greatest for seeds treated as in F, G or H?

Tick the correct box

0–4 days ☐   4–8 days ☐   8–12 days ☐

12–16 days ☐   16–20 days ☐   20–24 days ☐   **(1)**

(f) Calculate the difference between the number of days taken for 50% germination success to be reached in Treatments F and G.

Space for calculation

Number of days _____   **(1)**

(g) The abscisic acid concentration in seed coats has an effect on dormancy. What evidence supports this statement?

_____

_____   **(1)**

(h) Abscisic acid is present in seed coats. What evidence is there that chilling causes a loss of abscisic acid from seeds coats?

_____

_____   **(1)**

(i) Predict the percentage germination in Treatment G after 28 days.

Prediction _____ %   **(1)**

SECTION C

Both questions in this section should be attempted.
Note that each question contains a choice.

Questions 17 and 18 should be attempted on the blank pages which follow.

Supplementary sheets, if required, may be obtained from the invigilator.

Labelled diagrams may be used where appropriate.

Marks

17. Answer **either** A **or** B.

    A. Give an account of membranes under the following headings:

        (*a*) structure of membranes;   3

        (*b*) entry of materials into the cell;   7

        (*c*) protein transport and secretion.   5

                                                                      (15)

    OR

    B. The following techniques may be used to produce varieties of organisms. Give an account of these techniques and their benefits:

        (*a*) genetic engineering of bacteria to manufacture human protein;   7

        (*b*) somatic fusion in plants;   4

        (*c*) selective breeding.   4

                                                                      (15)

18. Answer **either** A **or** B.

    A. Give an account of natural selection and the evolution of new species.   **(15)**

    OR

    B. Give an account of the light and dark stages of photosynthesis.   **(15)**

[END OF QUESTION PAPER]

SCOTTISH
CERTIFICATE OF
EDUCATION
1998

MONDAY, 11 MAY
9.30 AM – 12.00 NOON

BIOLOGY
HIGHER GRADE
Paper II

1 (a) All questions should be attempted.

(b) It should be noted that questions 17 and 18 each contain a choice.

2 The questions may be answered in any order but all answers are to be written in the spaces provided in this answer book, and must be written clearly and legibly in ink.

3 Additional space for answers and rough work will be found at the end of the book. If further space is required, supplementary sheets may be obtained from the invigilator and should be inserted inside the front cover of this booklet.

4 The numbers of questions must be clearly inserted with any answers written in the additional space.

5 Rough work, if any should be necessary, should be written in this booklet and then scored through when the fair copy has been written.

6 Before leaving the examination room, you must give this book to the invigilator. If you do not, you may lose all the marks for this paper.

SECTION A

Answer ALL questions in this section.

Marks

1. The diagrams below show a chloroplast and an outline of the Calvin cycle.

Chloroplast　　　　　　　**Outline of Calvin cycle**

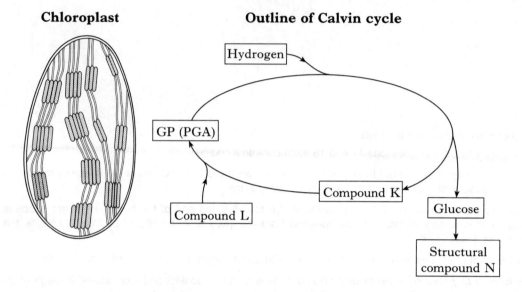

(a) On the diagram of the chloroplast, **put an X** to show where the Calvin cycle takes place.　　**(1)**

(b) Name compounds K and L in the Calvin cycle.

Compound K _____

Compound L _____　　**(1)**

(c) Complete the following sentence by **<u>underlining</u> one** of the words in each group.

Structural compound N is $\left\{\begin{array}{l}\text{starch}\\ \text{cellulose}\\ \text{glycogen}\end{array}\right\}$ which is deposited in the cell

wall as $\left\{\begin{array}{l}\text{fibres.}\\ \text{globules.}\\ \text{grains.}\end{array}\right\}$ The cell wall is $\left\{\begin{array}{l}\text{permeable}\\ \text{impermeable}\\ \text{selectively permeable}\end{array}\right\}$ to water.　　**(2)**

Marks

(*d*) The hydrogen required for the Calvin cycle comes from the light stage of photosynthesis.

(i) Name the source of the hydrogen.

Source _____ **(1)**

(ii) Name the compound responsible for hydrogen transfer to the Calvin cycle.

Compound _____ **(1)**

(iii) Name one other product of the light stage which is required for the Calvin cycle.

Product _____ **(1)**

Marks

2. (*a*) Two types of ribonucleic acid (RNA) are involved in protein synthesis. Complete the blanks in the following table to show the types of RNA, their functions and location(s) in a cell.

Type of RNA	Function	Location(s)
		1. Cytoplasm 2. _____
	Transfers amino acids to ribosomes	Cytoplasm

(2)

(*b*) The m-RNA codons for some amino acids are given in the table below.

Codon	Amino acid
AGA	arginine
CUC	leucine
ACA	threonine

State the order of bases on the deoxyribonucleic acid (DNA) which would result in the assembly of the following amino acid sequence.

leucine—threonine—arginine

_____ **(1)**

(*c*) Describe the function of the rough endoplasmic reticulum and the Golgi apparatus in a cell.

Rough endoplasmic reticulum _____

_____ **(1)**

Golgi apparatus _____

_____ **(1)**

Marks

3. (a) An investigation into an antibody is summarised below.

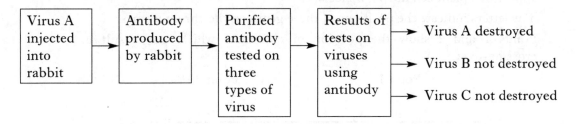

(i) Name the cells in the rabbit which produce antibodies.

_____ **(1)**

(ii) Specific antibodies are formed in response to foreign antigens. How do the results of the investigation support this statement?

_____

_____

_____ **(2)**

(b) The list below refers to structures and compounds involved in the defence of organisms.

List	Letter
antibody	A
cyanide	B
lymphocyte	C
lysosome	D
nicotine	E
resin	F
tannin	G

Use the letters from the list to identify:

(i) the cell organelle associated with phagocytosis;

Letter _____ **(1)**

(ii) the **four** chemical substances produced by plants for their defence.

Letters _____ **(1)**

Marks

4. The diagram below represents a pair of homologous chromosomes as they appear in a plant cell during meiosis.

The letters indicate the positions of three genes on the chromosomes.

Points 1 and 2 show the position of chiasmata which may result in crossing-over.

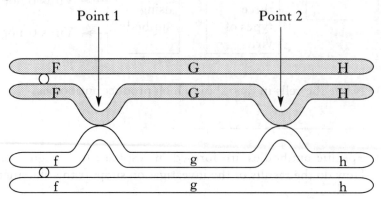

(a) State a location in a flowering plant where cells undergo meiosis.

Location _____ **(1)**

(b) Meiosis in the cell described above produces gametes with various genotypes. The grid below shows these genotypes.

1 FGH	2 FGh	3 FgH	4 Fgh
5 fgh	6 fgH	7 fGh	8 fGH

Use the **numbers** from the grid to answer the following questions.

(i) Which **two** genotypes are **not** recombinants?

Numbers _____ and _____ **(1)**

(ii) Which **two** genotypes result from crossing-over at Point 1 only?

Numbers _____ and _____ **(1)**

(iii) Which **two** genotypes result from crossing-over at both Points 1 and 2?

Numbers _____ and _____ **(1)**

Marks

(c) Between which two genes on the chromosomes shown opposite would the highest frequency of recombination take place?

Tick the correct box.

Between F and G ☐   Between G and H ☐   Between F and H ☐   **(1)**

(d) Apart from crossing-over and mutation, state **one** other source of genetic variation which occurs during meiosis.

_____   **(1)**

Marks

5. (*a*) Tetracycline is an antibiotic which kills bacteria.

Strains of bacteria which are resistant to tetracycline have evolved.

The information below shows stages in the evolution of tetracycline resistance in bacteria.

Stage 1	Original population of bacteria

Stage 2	A mutant bacterium arose in the population. The mutant had an abnormal membrane protein which prevented tetracycline entering the cell

Stage 3	The population of bacteria was exposed to tetracycline

Stage 4	Natural selection led to the evolution of a resistant population of bacteria

(i) Explain how gene mutation would lead to the production of the abnormal membrane protein described in *Stage 2*.

_____

_____

_____   **(2)**

(ii) Explain how tetracycline acts as an agent for natural selection and how this leads to the evolution of a resistant population of bacteria.

_____

_____

_____

_____   **(2)**

(*b*) Name an agent which causes gene mutations.

_____   **(1)**

6. The maps below show the positions of land masses in the Southern Hemisphere at three times in the past.

The chart below shows the stages in the evolution of three species of flightless birds in the land masses of the Southern Hemisphere following isolation.

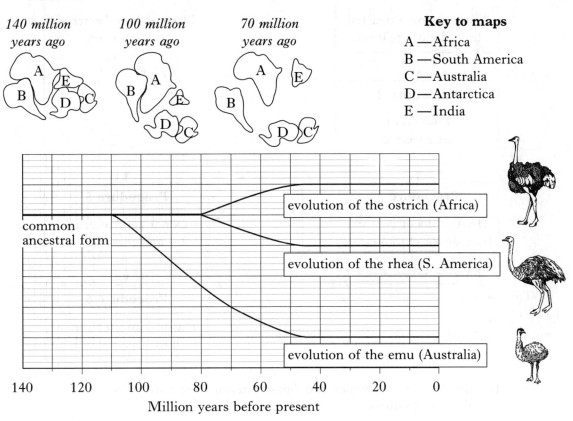

140 million years ago   *100 million years ago*   *70 million years ago*

Key to maps

A — Africa
B — South America
C — Australia
D — Antarctica
E — India

common ancestral form

evolution of the ostrich (Africa)

evolution of the rhea (S. America)

evolution of the emu (Australia)

140   120   100   80   60   40   20   0

Million years before present

Marks

(a) What type of isolation mechanism is shown by the maps?

_____

(1)

(b) From the chart, how many millions of years separated the isolation of the emu from the isolation of the ostrich from the rhea?

Number _____

(1)

(c) Use the information in the maps and chart to account for the evolution of these three species from a common ancestral form.

_____

_____

_____

_____

(2)

Marks

7. The information below shows some of the procedures used in genetic engineering to insert a gene from a donor organism into a bacterial cell.

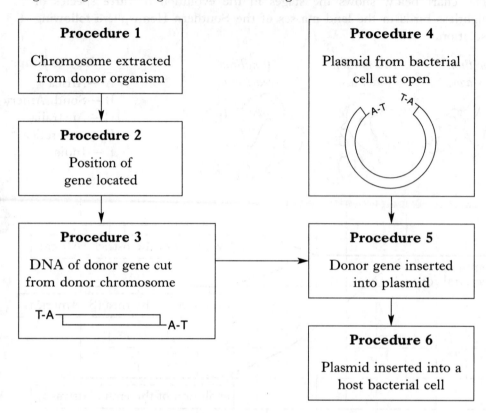

(a) Endonuclease enzymes and ligase enzymes are used in some of the above procedures.

   (i) Identify the **two** procedures which use **endonuclease** enzymes.

   Procedures _____ and _____   **(1)**

   (ii) Identify the procedure which uses **ligase** enzymes.

   Procedure _____   **(1)**

(b) Name a technique which could be used in **Procedure 2** to locate the position of a gene on a chromosome.

   Technique _____   **(1)**

(c) Which feature, shown in the procedures, allows insertion of the isolated donor gene into the open plasmid during **Procedure 5**?

   _____

   _____

   _____   **(1)**

Marks

8. (*a*) Complete the table below to show the effects of macro-element deficiency in plants.

Element omitted	Symptoms of deficiency	Importance of element
	1. Overall growth reduced 2. Early death of older leaves	Membrane transport
Phosphorus	1. Overall growth reduced 2. Leaf bases red	
Magnesium	1. Overall growth reduced 2. _____ _____	

(4)

(*b*) The following graph compares the uptake of nitrate ions by barley roots in the presence and absence of oxygen.

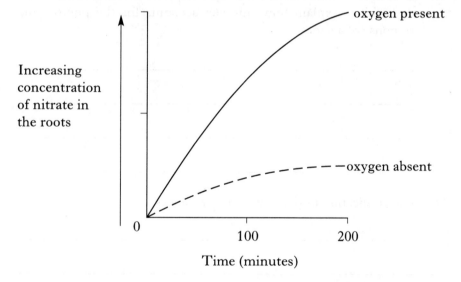

Increasing concentration of nitrate in the roots

oxygen present

oxygen absent

0   100   200

Time (minutes)

Account for the effect of oxygen on the uptake of nitrate by the roots.

_____

_____

_____

_____

(2)

Marks

9. (*a*) The diagrams below show the results of experiments on the growth of plant shoots.

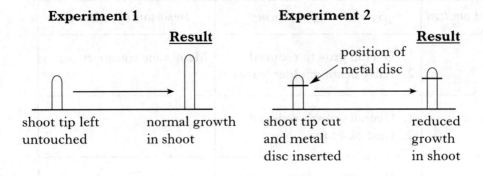

Experiment 1

Result

shoot tip left untouched   normal growth in shoot

Experiment 2

Result

position of metal disc

shoot tip cut and metal disc inserted   reduced growth in shoot

    (i) How do the results support the statement that the shoot tip is the site of production of IAA?

_____

_____ **(1)**

    (ii) If a shoot is illuminated from one side, IAA is transported to the dark side. Explain how this fact accounts for the phototropic response of a shoot.

_____

_____

_____

_____ **(2)**

(*b*) Give one application of plant growth substances.

_____

_____ **(1)**

Marks

10. (*a*) The flow chart below represents part of the homeostatic control of blood glucose concentration in a human.

Increase in
blood glucose → Detected by → Corrective mechanism 1
concentration   receptor cells

Normal blood glucose concentration

Normal blood glucose concentration

Decrease in
blood glucose → Detected by → Corrective mechanism 2
concentration   receptor cells

(i) State the location of the receptor cells.

_____ **(1)**

(ii) Complete each box in the table below by inserting the word **increases** or **decreases**.

	Insulin concentration	*Glucagon concentration*
During corrective mechanism 2		

(1)

(iii) In corrective mechanism 1, glucose is removed from the blood and is converted into a storage carbohydrate.

Name the storage carbohydrate and name the organ in which it is stored.

Name _____

Organ _____ **(2)**

(*b*) The following statements refer to endotherms and ectotherms.

Tick the appropriate box to indicate if the statement is **true** for endotherms **and/or** ectotherms.

	Endotherm	*Ectotherm*
Can only regulate body temperature through change in behaviour		
Body temperature held constant independently of external environmental temperature		
Metabolic rate increases with increase in external temperature		

(2)

Marks

11. When an area of land is cleared of vegetation and left to be colonised, succession takes place.

The table below gives information about two plant species which grow at different stages in this succession. Other changes which take place during this succession are also shown.

	Stage in succession	
	Early succession	*Late succession (climax)*
Plant species	Rosebay willowherb	Beech
Information on plant species	Fast growing with bright pink flowers. The fruit releases many, light, hairy seeds.	A large tree reaching 30 metres in height. Leaves form a dense canopy. The fruit is a nut and is eaten by animals such as rabbits.
Biomass of community	---------- increasing/decreasing/does not change --------➤	
Complexity of food webs	---------- increasing/decreasing/does not change --------➤	

(*a*) Show the effect of succession on biomass and complexity of food webs by underlining **increasing**, **decreasing** or **does not change** in the table above.

(1)

(*b*) (i) Select a feature of rosebay willowherb and explain how this feature helps the plant to be an efficient early coloniser.

Feature _____

Explanation _____

_____

_____

(1)

(ii) Select a feature of beech and explain how this feature helps to prevent other plant species from succeeding it.

Feature _____

Explanation _____

_____

_____

(2)

Marks

12. Freshwater and salt water bony fish have adaptations which maintain the water concentration of their body fluids at a constant value.

(a) Name the process responsible for the movement of water into and/or out of cells.

Process ——————————————— **(1)**

(b) **Underline** the correct answer in the statement below.

Salt water fish are $\left\{\begin{array}{l}\text{hypertonic}\\\text{hypotonic}\\\text{isotonic}\end{array}\right\}$ to their environment. **(1)**

(c) In the list below, **tick** (✓) the **four** boxes which show features of osmoregulation in freshwater fish.

List	Tick (✓)
High glomerular filtration rate	
Few glomeruli present	
Active uptake of salts at gills	
Concentrated urine produced	
Low urine volume produced	
Low glomerular filtration rate	
Dilute urine produced	
Many glomeruli present	
Active secretion of salts at gills	

(2)

Marks

13. The grid below refers to types of behaviour which birds use to obtain food and to defend themselves.

A Cooperative hunting	B Behaviour which reduces interspecific competition	C Dominance hierarchy
D Territorial behaviour	E Social defence	F Behaviour which reduces intraspecific competition

The table below contains some examples of bird behaviour.

Use the letters from the grid to identify the type of behaviour described in each example.

Each letter can be used **once**, **more than once**, or **not at all**.

Example of behaviour	*Letter(s)*
Pelicans, searching for food, form a circle around fish, then dip their beaks into the water simultaneously.	
The Sooty Tern feeds on larger fish than other species of Tern which live in the same area.	
In a population of Great Tits, birds with the widest stripe on their breast feed first when food is scarce.	

(3)

SECTION B

Answer ALL questions in this section.

14. Animals may show escape responses when first exposed to certain stimuli. These responses may alter as a result of habituation.

In a laboratory experiment, twenty snails of the same species were observed for 20 minutes. During part of the experiment, a stimulus (X) was applied to the animals at 2 minute intervals. Throughout the experiment the number of animals showing an escape response was noted at 2 minute intervals.

The results are shown below.

Time (minutes)	0	2	4	6	8	10	12	14	16	18	20
Stimulus applied	none	none	none	none	stimulus X	stimulus X	stimulus X	stimulus X	stimulus X	stimulus X	stimulus X
Number showing escape response	0	0	0	0	20	19	20	14	8	0	0

(*a*) Why was it good experimental procedure to wait for 8 minutes before applying the stimulus for the first time?

_____

_____ **(1)**

(*b*) Results can be invalid if the stimulus is not applied correctly.

State **one** precaution which should be taken when applying the stimulus.

_____

_____ **(1)**

(*c*) Explain how the results indicate that habituation has occurred in this experiment.

_____

_____ **(1)**

(*d*) Habituation is a **short term** change in behaviour.

Describe an experiment which would demonstrate this feature of habituation.

_____

_____ **(2)**

Marks

In a further experiment, a different stimulus (Y) was applied instead of stimulus X.

The results are shown below.

Time (minutes)	0	2	4	6	8	10	12	14	16	18	20
Stimulus applied	none	none	none	none	stimulus Y	stimulus Y	stimulus Y	stimulus Y	stimulus Y	stimulus Y	stimulus Y
Number showing escape response	0	0	0	0	20	19	19	20	20	19	20

(*e*) State how the response of the animals to stimulus Y differs from that to stimulus X. Suggest why this difference may be important for animals in coping with danger.

Difference in response ______________________________

______________________________ **(1)**

Importance ______________________________

______________________________ **(1)**

In another experiment, stimulus X was applied, but a third stimulus (Z) was substituted from 18 minutes.

The results are shown below.

Time (minutes)	0	2	4	6	8	10	12	14	16	18	20	22	24
Stimulus applied	none	none	none	none	stimulus X	stimulus X	stimulus X	stimulus X	stimulus X	stimulus Z	stimulus Z	stimulus Z	stimulus Z
Number showing escape response	0	0	0	0	20	19	8	1	0	19	19	7	0

(*f*) Identify a feature of habituation which is shown by the results of the third experiment, and explain how this feature would be beneficial to animals.

Feature ______________________________

______________________________ **(1)**

Explanation ______________________________

______________________________ **(1)**

Marks

15. Between 1970 and 1995, thousands of kilometres of hedgerow were cut down to make larger fields for growing and harvesting crops.

The table below shows the length of hedgerow present in the United Kingdom at four dates.

Year	Length of hedgerow (km)
1970	515 000
1984	412 000
1990	371 000
1995	309 000

(a) Calculate the percentage reduction in the length of hedgerow between 1970 and 1984.

Space for calculation

Answer _____ %   **(1)**

(b) During which period did half of the total hedgerow destruction take place?

Tick the correct box.

1970–1984 ☐   1984–1990 ☐   1990–1995 ☐   **(1)**

The graph below shows how crop yield and several abiotic factors change with distance from a hedge.

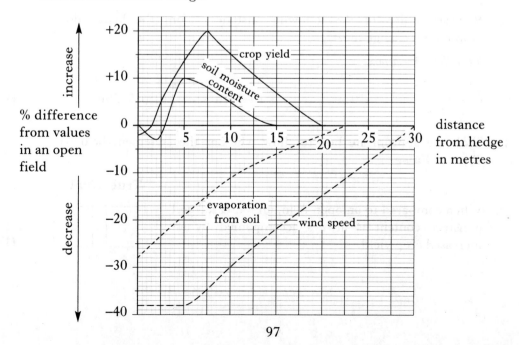

97

Marks

(c) What is the evidence that wind speed is not the only factor which affects evaporation of water from the soil?

_____

_____ **(1)**

(d) The yield of the crop in the open field is $0.8\,kg/m^2$.
Calculate the crop yield at a distance of 7.5 metres from the hedge.

Space for calculation

Crop yield _____ kg/m^2 **(1)**

(e) Up to what distance from the hedge does its presence lead to a decrease in crop yield when compared to the values in an open field?

Distance _____ metres **(1)**

(f) Describe how soil moisture content changes with distance from the hedge.

_____

_____

_____

_____ **(2)**

(g) Between which of the following distances from the hedge does evaporation change most?

Tick the correct box.

0–5 m ☐   5–10 m ☐   10–15 m ☐   15–20m ☐   **(1)**

(h) **Tick** the appropriate box to show whether the statement below is **True** or **False**.

True   **False**

When compared to an open field, an increased soil moisture content is always accompanied by an increased crop yield.   ☐   ☐   **(1)**

Marks

16. The population density of red deer in Scotland varies from area to area.

Population density affects survival, growth and fertility of the deer.

The bar chart shows the results of population surveys in areas with low population densities and areas with high population densities.

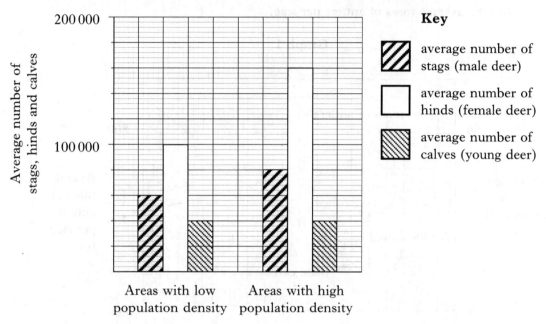

(a) The ratio of hinds to stags in areas with low population densities was 5 hinds : 3 stags.

Calculate the simplest whole number ratio of hinds to stags in areas with high population densities.

Space for calculation

Ratio _____ Hinds : _____ Stags   **(1)**

Marks

Graph 1 below shows the results of a study in which areas with different population densities were compared.

In each area, measurements were made of the following:

(i) the percentage of calves which died;

(ii) the average mass of antlers per stag.

Graph 1

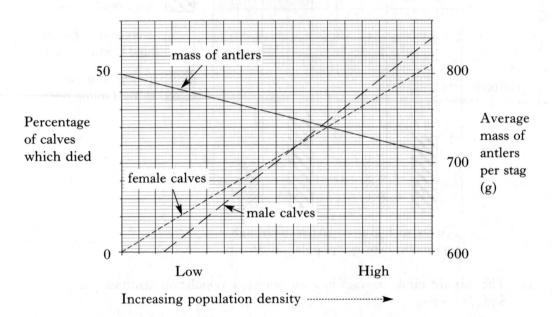

(b) What evidence from Graph 1 suggests that growth of the deer is affected by population density?

_____

_____ **(1)**

(c) The proportion of stags in areas with high population densities is less than in areas with low population densities.

Explain how the data in Graph 1 could account for this difference.

_____

_____

_____ **(1)**

Marks

Graph 2 shows a comparison of fertility in deer populations in eight different locations in Scotland.

In each location, measurements were made of the population density and the percentage of one-year-old hinds which were pregnant.

Populations living in woodland and open hill areas were compared.

Graph 2

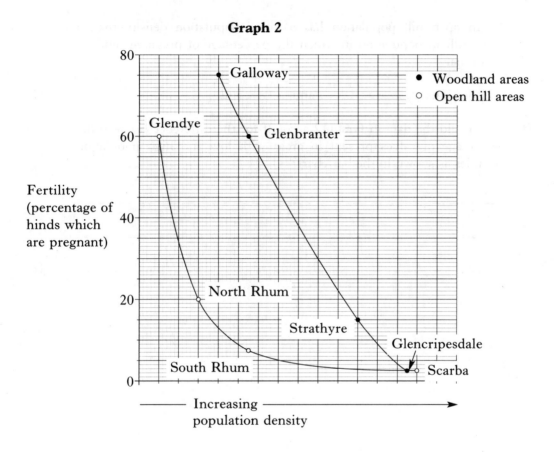

(d) Use the data in Graph 2 to describe the effects of increasing population density on fertility in open hill areas.

_____

_____

_____

_____　　**(2)**

Marks

(e) Name the **two** locations shown in Graph 2 which should be compared to support each of the following statements.

1. A woodland population has a higher percentage of pregnant hinds than an open hill population of a similar density.

   Locations _____ and _____   **(1)**

2. An open hill population has a lower population density than a woodland population in which the percentage of pregnant hinds is similar.

   Locations _____ and _____   **(1)**

(f) How could the data in Graph 2 account for the difference between the percentage of calves present in areas with low and high population densities as shown in the bar chart?

_____

_____

_____   **(2)**

SECTION C

Both questions in this section should be attempted.
Note that each question contains a choice.

Questions 17 and 18 should be attempted on the blank pages which follow.

Supplementary sheets, if required, may be obtained from the invigilator.

Labelled diagrams may be used where appropriate.

Marks

17. Answer **either** A **or** B.

    A. Write notes on each of the following:

        (*a*)   position and activity of meristems;   **5**

        (*b*)   formation of annual rings;   **5**

        (*c*)   the plant growth substance gibberellic acid.   **5**

    OR   **(15)**

    B. Write notes on each of the following:

        (*a*)   the Jacob-Monod hypothesis;   **6**

        (*b*)   phenylketonuria;   **4**

        (*c*)   the role of the pituitary gland.   **5**

        (15)

18. Answer **either** A **or** B.

    A. Give an account of the production of ATP in cell respiration.   **(15)**

    OR

    B. Give an account of the transpiration stream and explain how xerophytes are adapted to reduce water loss by transpiration.   **(15)**

[END OF QUESTION PAPER]

SCOTTISH
CERTIFICATE OF
EDUCATION
1999

WEDNESDAY, 12 MAY
9.00 AM – 11.30 AM

BIOLOGY
HIGHER GRADE
Paper II

1 (a) All questions should be attempted.

(b) It should be noted that questions 16 and 17 each contain a choice.

2 The questions may be answered in any order but all answers are to be written in the spaces provided in this answer book, and must be written clearly and legibly in ink.

3 Additional space for answers and rough work will be found at the end of the book. If further space is required, supplementary sheets may be obtained from the invigilator and should be inserted inside the front cover of this booklet.

4 The numbers of questions must be clearly inserted with any answers written in the additional space.

5 Rough work, if any should be necessary, should be written in this booklet and then scored through when the fair copy has been written.

6 Before leaving the examination room, you must give this book to the invigilator. If you do not, you may lose all the marks for this paper.

SECTION A

Answer ALL questions in this section.

1. (a) The diagram below illustrates movement of molecules of substances A and B across the membrane of a human kidney cell.

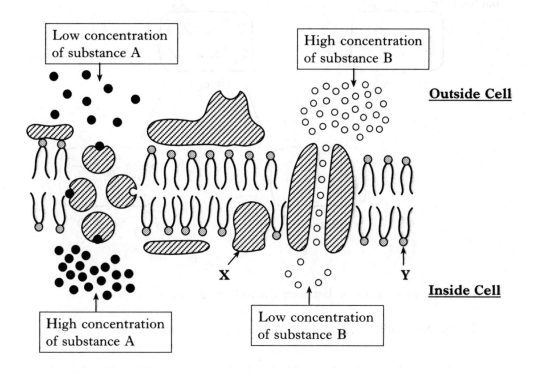

Marks

    (i) Identify membrane components X and Y.

        X _____

        Y _____   **(1)**

    (ii) Name the methods by which substances A and B **enter** the cell.

        Method for A _____

        Method for B _____   **(2)**

(b) Some membrane molecules act as antigens. When a kidney is transplanted into a patient, antibodies are produced in response to these antigens. Name the cells responsible for antibody production.

_____   **(1)**

Marks

2. The diagram below shows invasion of a bacterial cell by a virus.

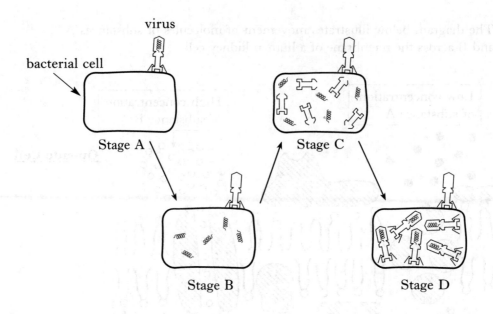

(a) Describe what happens between the following stages.

1   Stages A and B _____

_____

_____ **(1)**

2   Stages B and C _____

_____

_____ **(1)**

(b) State what takes place after stage D is complete.

_____

_____ **(1)**

Marks

3. (*a*) The diagram below shows an outline of some stages in cell respiration.

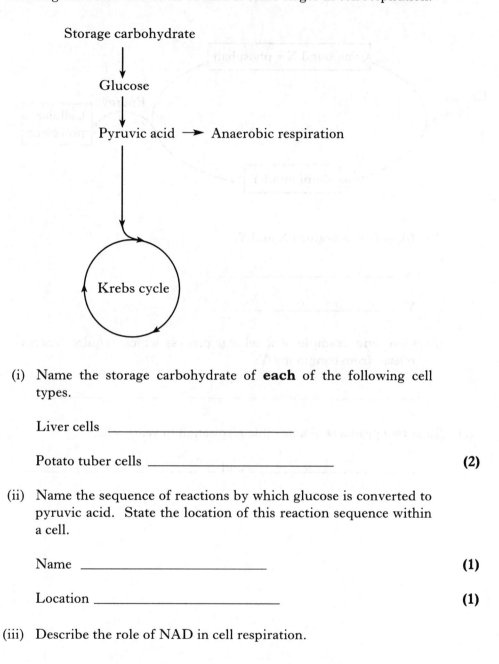

Storage carbohydrate

Glucose

Pyruvic acid → Anaerobic respiration

Krebs cycle

(i) Name the storage carbohydrate of **each** of the following cell types.

Liver cells _____

Potato tuber cells _____   **(2)**

(ii) Name the sequence of reactions by which glucose is converted to pyruvic acid. State the location of this reaction sequence within a cell.

Name _____   **(1)**

Location _____   **(1)**

(iii) Describe the role of NAD in cell respiration.

_____

_____   **(1)**

Marks

(b) The diagram below shows energy transfer within a cell.

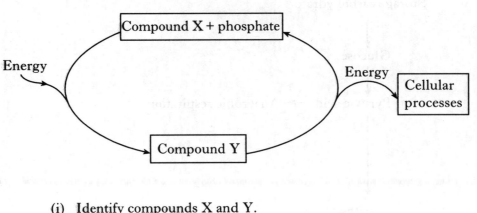

(i) Identify compounds X and Y.

X _____

Y _____ **(1)**

(ii) Give **one** example of a cellular process which requires energy release from compound Y.

_____ **(1)**

(c) Name **two** products of anaerobic respiration in yeast cells.

_____ and _____ **(1)**

4. (*a*) The diagram below shows two human karyotypes.

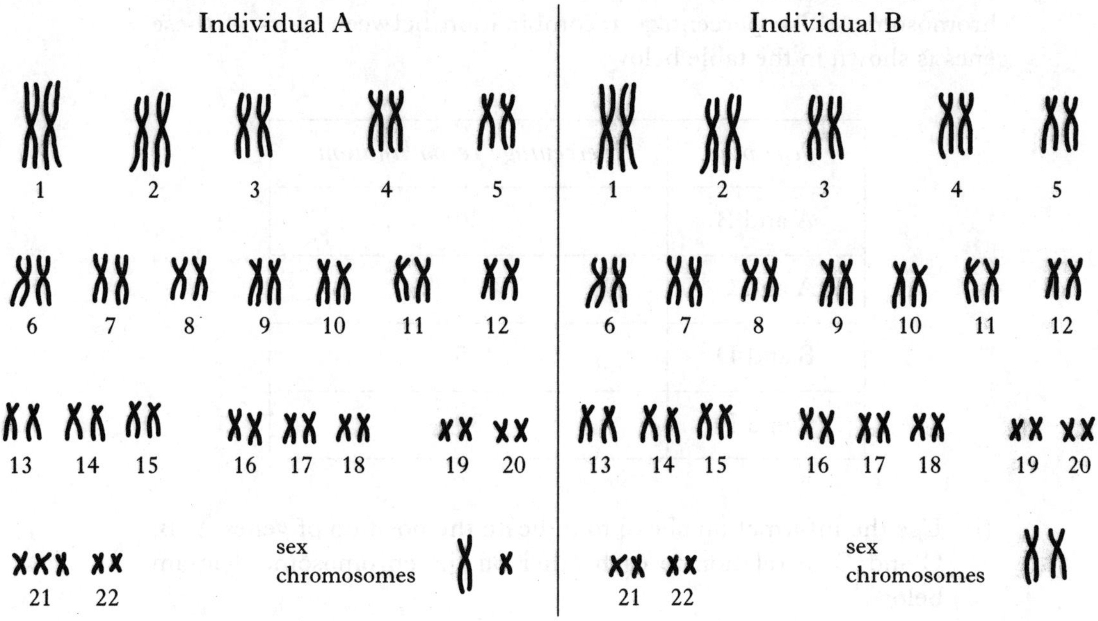

Individual A   Individual B

Marks

(i) State which individual is male and give a reason for your answer.

Individual _____

Reason _____

_____ **(1)**

(ii) Karyotype A shows a mutation which affects chromosome set 21. Complete the table below to name the organ in which this mutation may occur and the inherited disorder which results.

Organ where mutation occurs	
Name of inherited disorder	

(2)

(iii) A pair of homologous chromosomes are the same size and shape. State **one** other feature which allows a pair of chromosomes to be described as homologous.

_____

_____ **(1)**

Marks

(*b*) In a certain organism, genes A, B, C and D are located on the same chromosome. The percentage recombination between pairs of these genes is shown in the table below.

Gene pair	Percentage recombination
A and B	20
A and C	5
B and D	5
C and D	30

(i) Use the information above to indicate the position of genes A, B, C and D in relation to each other on the chromosome diagram below.

Space for calculation

Chromosome _____ **(1)**

(ii) How can recombination lead to variation in the genotype of gametes?

_____

_____ **(1)**

Marks

5. The diagram below represents part of a DNA molecule during replication.

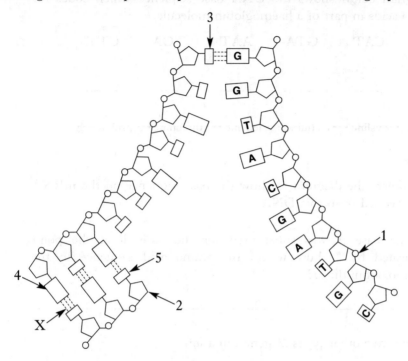

(*a*) What is the first event in the process of DNA replication?

_____ **(1)**

(*b*) From the diagram, identify components 1 and 2.

1 _____

2 _____ **(1)**

(*c*) Identify the type of bond labelled as X on the diagram.

_____ **(1)**

(*d*) From the diagram, identify each of the bases 3, 4 and 5.

Base 3 _____

Base 4 _____

Base 5 _____ **(2)**

Marks

6. (*a*) The diagram below shows the DNA base sequence which codes for the amino acids in part of a haemoglobin molecule.

DNA   CAT   GTA   AAT   TGA   CTT*

mRNA   ____   ____   ____   ____   ____

Amino acid   --- valine --- histidine --- leucine --- threonine --- proline ---
sequence

(i) Complete the diagram to show the base sequence on the mRNA synthesised from this DNA.   **(1)**

(ii) If the base A was substituted for the base T at the point indicated by *, how would the amino acid sequence in the haemoglobin differ?

_____   **(1)**

(iii) Name **two** other types of gene mutation.

_____ and _____   **(1)**

(iv) What name is given to a triplet of bases found on a strand of mRNA which codes for one amino acid?

_____   **(1)**

(*b*) The diagram below represents part of a metabolic pathway within a cell.

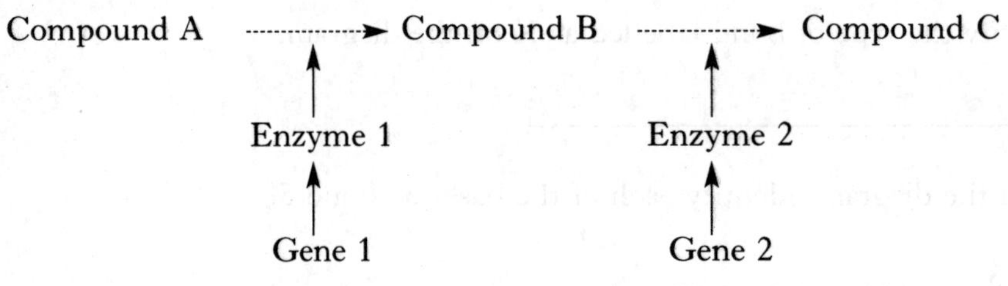

Compound A   ----------►   Compound B   ----------►   Compound C

                    ↑   ↑
              Enzyme 1   Enzyme 2
                    ↑   ↑
               Gene 1   Gene 2

Which compound could accumulate within the cell if a mutation occurred in Gene 1?

Explain your answer.

Compound   _____

Explanation   _____

_____   **(2)**

7. (*a*) Somatic fusion is a technique used in plant breeding.

The flow diagram below shows an outline of this process.

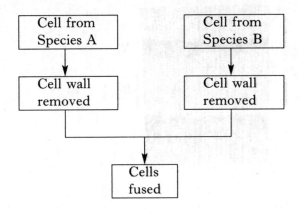

(i) Name the enzyme used to remove cell walls.

_____ **(1)**

(ii) What term is used to describe a cell formed after the cell wall is removed?

_____ **(1)**

(iii) What problem in plant breeding is overcome by somatic fusion?

_____

_____ **(1)**

(*b*) Human insulin can be produced by genetic engineering techniques. Enzymes are used at different stages in this process.

Complete the blanks in the following table to show the enzymes used and their function.

Name of enzyme	Function in process
Endonuclease (restriction enzyme)	
	Joins DNA molecules

(1)

(*c*) Name a technique used to locate genes on a human chromosome.

_____ **(1)**

Marks

8. The diagram below shows a cross section through xylem vessels forming annual rings in the trunk of a tree.

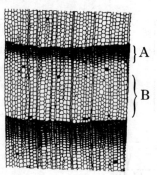

(a) Which letter identifies the vessels formed in the spring?

Give a reason for your answer.

Letter _____

Reason _____

_____ **(1)**

(b) The formation of xylem vessels follows the pattern shown below.

mitosis in meristematic cells ⟶ cell differentiation ⟶ formation of xylem vessels

(i) Name the meristematic tissue responsible for the formation of annual rings.

_____ **(1)**

(ii) Describe **two** structural changes that take place as a cell becomes differentiated to form part of a xylem vessel.

1. _____

_____ **(1)**

2. _____

_____ **(1)**

(c) Other cells in the stem become differentiated to form phloem tissue. How can the formation of different tissues during development be explained in terms of gene activity?

_____

_____ **(1)**

9. The graph below shows the annual variation in the light intensity at the canopy and ground layers in a deciduous woodland.

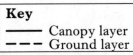

Canopy layer

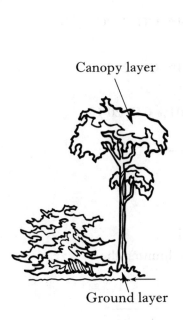

Ground layer

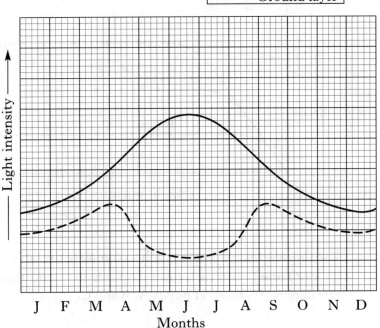

(a) Account for the changes in light intensity at the ground layer throughout the year.

Marks

_____

_____

_____ **(2)**

(b) (i) Complete the following sentences by **underlining** one of the words in each group.

Ground layer plants are $\begin{Bmatrix} \text{sun} \\ \text{shade} \end{Bmatrix}$ plants. Compared to canopy plants, their net gain of carbohydrate is $\begin{Bmatrix} \text{greater} \\ \text{less} \end{Bmatrix}$ at lower light intensities. **(1)**

(ii) What name is given to the light intensity at which the rate of photosynthesis is equal to the rate of respiration?

_____ **(1)**

(c) The graph below shows the rate of photosynthesis under different conditions.

Marks

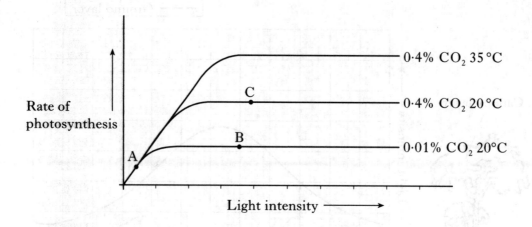

Use the information in the graph to complete the table below.

Tick (✓) **one** box in each row to indicate the factor that is limiting the rate of photosynthesis at points A, B and C.

Graph point	Temperature	Light intensity	CO_2 concentration
A			
B			
C			

(2)

Marks

10. The flow diagram below shows the effects of two hormones on growth and development in the human body.

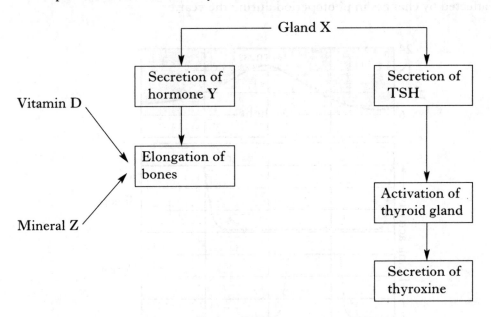

(a) Name gland X, hormone Y and mineral Z.

Gland X _____

Hormone Y _____

Mineral Z _____   **(2)**

(b) State the role of vitamin D in promoting bone growth.

_____

_____   **(1)**

(c) The control of thyroid activity is an example of negative feedback control. Given this information, predict the effect of an increase in thyroxine on gland X.

_____

_____   **(1)**

11. (*a*) Breeding behaviour in certain mammals is influenced by photoperiod. *Marks*
The graphs below show how the mating activity of sheep and ferrets is
affected by changes in photoperiod during the year.

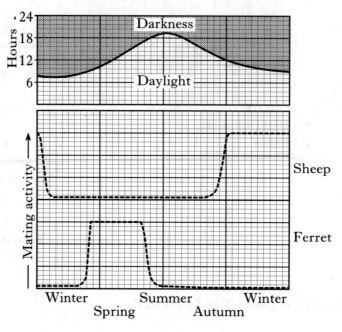

(i) Describe the relationship between photoperiod and the onset of
mating activity in sheep.

_____

_____ **(1)**

(ii) It is several months after mating before lambs are born whereas
ferrets give birth only weeks after mating.

Suggest how the timing of mating activity is advantageous for
survival of the young of both species.

_____

_____

_____ **(2)**

(*b*) Under certain conditions the growing stem of a plant may show
etiolation.

(i) Under what environmental conditions would etiolation occur?

_____ **(1)**

(ii) Describe **one** way in which an etiolated plant differs in
appearance from a normal plant.

_____ **(1)**

Marks

12. The grid below shows some of the factors that are involved in the flow of water through a plant.

A	B	C
turgor of guard cells	adhesive forces	diffusion along a concentration gradient
D	E	F
cohesive forces	osmosis	light intensity

(a) **Use the letters** from the boxes above to answer the following question. Which **two** factors contribute to movement of water molecules as a continuous column within xylem vessels?

Letters _____ and _____ **(1)**

(b) Describe how changes in factor A can affect the rate of transpiration.

_____

_____

_____ **(2)**

(c) Stomata open as conditions change from dark to light.

Why is this important to plants?

_____

_____ **(1)**

(d) State **one** benefit of transpiration to plants.

_____

_____ **(1)**

SECTION B

Answer ALL questions in this section.

13. The graphs below show some of the changes that take place in fully-grown leaves during a 10 day period in autumn.

Graph 1 shows changes in the mass of protein and chlorophyll per cm^2 of leaf area.

Graph 2 shows changes in the rate of photosynthesis, expressed as the rate of carbon dioxide uptake.

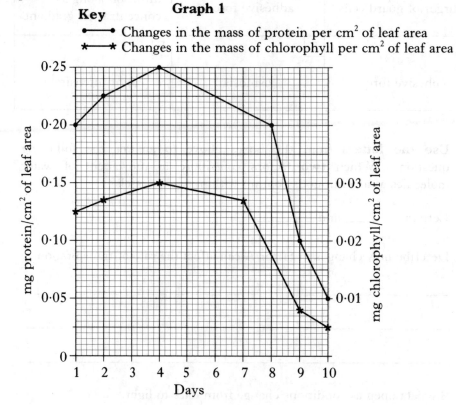

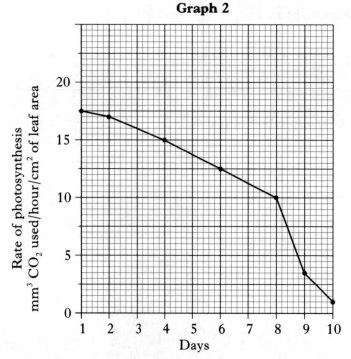

120

Marks

(*a*) From **Graphs 1** and **2**, what is the rate of photosynthesis when chlorophyll content is at a maximum?

_____ mm³ carbon dioxide used/hour/cm² of leaf area   **(1)**

(*b*) From **Graph 2**, calculate the volume of carbon dioxide used in 10 hours by a leaf with a surface area of 20 cm² on Day 6.
Space for calculation

_____ mm³   **(1)**

(*c*) Express the relationship between the masses of protein and chlorophyll per cm² of leaf area on Day 10 as a simple whole number ratio.
Space for calculation

_____ protein : _____ chlorophyll   **(1)**

(*d*) From **Graph 1**, calculate the percentage decrease in the mass of protein per cm² of leaf area from Day 4 to Day 9.
Space for calculation

_____ %   **(1)**

(*e*) In **Graph 1**, the protein content of leaves is expressed in mg of protein/cm² of leaf area.

Why could this method of expressing protein content **not** be used to give a fair comparison between the leaves of different plant species?

_____

_____   **(1)**

(*f*) Chlorophyll content of leaves may limit the rate of photosynthesis.
Use the data to comment on this statement.

_____

_____

_____   **(2)**

Marks

14. In humans, changes in the level of exercise bring about change in pulse rate and stroke volume (the volume of blood pumped from the heart in one heartbeat).

Graph 1 shows how pulse rate and stroke volume change with level of exercise.

Level of exercise is measured as rate of oxygen uptake.

Graph 1

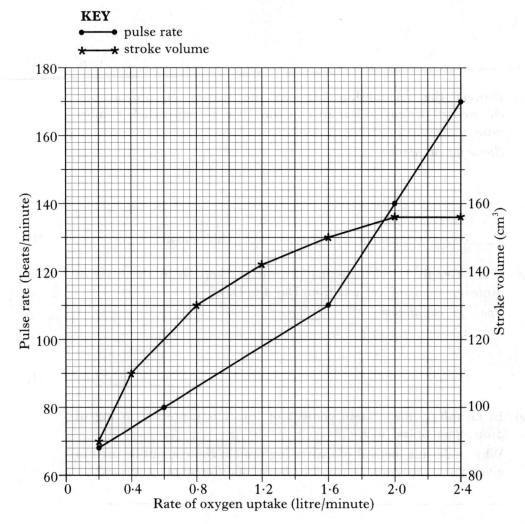

(a) What is the pulse rate and the stroke volume when the rate of oxygen uptake is 1·6 litres/minute?

Pulse rate _____ beats/minute

Stroke volume _____ cm³ **(1)**

Marks

(b) Calculate, in litres, the total volume of blood leaving the heart in one minute when the rate of oxygen uptake is 0·6 litres/minute.

Space for calculation

_____ litres **(1)**

(c) Calculate the increase in the rate of oxygen uptake when the pulse rate is increased from 70 beats/minute to 140 beats/minute.

Space for calculation

_____ litres oxygen/minute **(1)**

(d) What evidence from **Graph 1** would support the statement that there is a limit to the volume of blood that the heart can hold?

_____ **(1)**

Ventilation rate is calculated as the volume of air inhaled during one minute.

Graph 2 shows how ventilation rate changes with level of exercise.

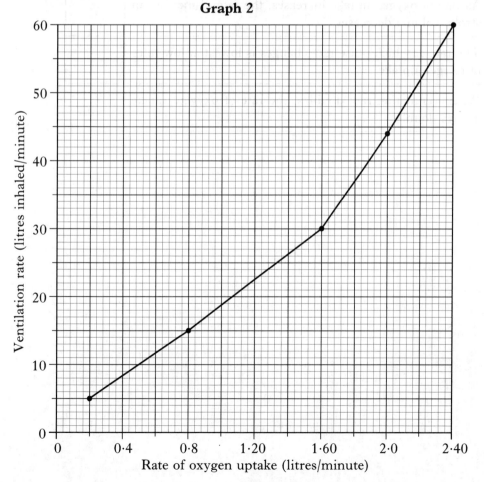

Graph 2

Marks

(e) Given that 20% of air is oxygen, calculate the volume of the oxygen inhaled per minute when the rate of oxygen uptake is 2·4 litres/minute.

Space for calculation

_____ litres **(1)**

(f) What additional information would be required to calculate the average volume of air taken in during each breath over one minute?

_____

_____ **(1)**

(g) Using information from **Graphs 1** and **2**, complete the table below by writing **T** if you think that the statement is **true** and **F** if you think that the statement is **false**.

Statement	*T or F*
As rate of oxygen uptake increases, the rate of increase in stroke volume decreases	
The rate at which pulse rate changes is lowest at low rates of oxygen uptake	
When ventilation rate doubles, the rate of oxygen uptake always doubles	

(3)

15. An investigation was carried out into the effect of different concentrations of indole-acetic acid (IAA) on growth in length of shoot tips.

Two 10 mm lengths of shoot tip tissue were immersed in solutions containing different concentrations of IAA.

A control experiment was set up with two 10 mm lengths of shoot tip tissue immersed in distilled water.

The results obtained are shown in the table below.

- Promotion (+) in growth is obtained when the growth is greater than in the control.

- Inhibition (−) in growth is obtained when the growth is less than in the control.

Concentration of IAA solution (molar)	Average length of shoot after 24 hours (mm)	Average difference in length compared to control (mm)
Control	**12·0**	
10^{-9}	12·0	0·0
10^{-8}	13·0	+1·0
10^{-7}	14·0	+2·0
10^{-6}	15·0	+3·0
10^{-5}	16·0	+4·0
10^{-4}	14·5	+2·5
10^{-3}	13·0	+1·0
10^{-2}	12·0	0·0
10^{-1}	11·0	−1·0

Increasing concentration

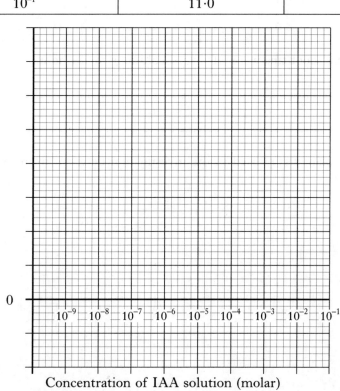

Concentration of IAA solution (molar)

Marks

(*a*) Complete the graph for the results obtained for average difference in length of shoot tips compared to control at different concentrations of IAA. **(2)**

(*b*) Identify the range of IAA concentrations that promote shoot growth.

From: _____ molar   to: _____ molar   **(1)**

(*c*) State **one** way in which the design of the experiment could be changed to increase the reliability of the results.

_____

_____ **(1)**

(*d*) Predict the effect of an IAA concentration of 10^{-10} molar on shoot growth.

_____

_____ **(1)**

(*e*) Explain why $-2\,mm$ would be the greatest value for inhibition of growth in shoots.

_____

_____ **(1)**

(*f*) Suggest why the shoot tips were left in the solutions for 24 hours.

_____

_____ **(1)**

(*g*) Why must the plant tissue used in the investigation be taken from the tip of the shoot?

_____

_____ **(1)**

Marks

(h) The concentration of IAA is varied in the investigation. Other variables, however, must be kept constant.

Name **two** variables that must be kept constant during the investigation.

1. _____

2. _____   **(2)**

(i) Describe the importance of the control experiment in this investigation.

_____

_____

_____   **(1)**

(j) Describe the effect of IAA on cells in the shoot tip.

_____

_____   **(1)**

SECTION C

Both questions in this section should be attempted.
Note that each question contains a choice.

Questions 16 and 17 should be attempted on the blank pages which follow.

Supplementary sheets, if required, may be obtained from the invigilator.

Labelled diagrams may be used where appropriate.

Marks

16. Answer **either** A **or** B.

   A. Write notes on each of the following:

(*a*)	the need to maintain body temperature within tolerable limits;	2
(*b*)	control of water concentration in the blood of a mammal;	6
(*c*)	control of blood glucose concentration in a mammal.	7

   (15)

   OR

   B. Write notes on each of the following:

(*a*)	foraging behaviour in animals;	4
(*b*)	social mechanisms in animals for obtaining food and for defence;	6
(*c*)	defence mechanisms in plants.	5

   (15)

17. Answer **either** A **or** B.

   A. Give an account of the process of photosynthesis. **(15)**

   OR

   B. Give an account of the role of isolation mechanisms and natural selection in the evolution of new species. **(15)**

[END OF QUESTION PAPER]

Printed by Bell & Bain Ltd., Glasgow, Scotland.